Undergraduate Lecture Notes in Physics

Series Editors

Neil Ashby, University of Colorado, Boulder, USA

William Brantley, Department of Physics, Furman University, Greenville, USA

Matthew Deady, Physics Program, Bard College, Annandale-on-Hudson, USA

Michael Fowler, Department of Physics, University of Virginia, Charlottesville, USA

Morten Hjorth-Jensen, Department of Physics, University of Oslo, Oslo, Norway

Michael Inglis, Department of Physical Sciences, SUNY Suffolk County Community College, Selden, USA

Barry Luokkala , Department of Physics, Carnegie Mellon University, Pittsburgh, USA

Antigone Marino, Fisica, University of Naples Federico II, Naples, Italy

Undergraduate Lecture Notes in Physics (ULNP) publishes authoritative texts covering topics throughout pure and applied physics. Each title in the series is suitable as a basis for undergraduate instruction, typically containing practice problems, worked examples, chapter summaries, and suggestions for further reading.

ULNP titles must provide at least one of the following:

- An exceptionally clear and concise treatment of a standard undergraduate subject.
- A solid undergraduate-level introduction to a graduate, advanced, or non-standard subject.
- A novel perspective or an unusual approach to teaching a subject.

ULNP especially encourages new, original, and idiosyncratic approaches to physics teaching at the undergraduate level.

The purpose of ULNP is to provide intriguing, absorbing books that will continue to be the reader's preferred reference throughout their academic career.

Reinhard Hentschke

A Concise Introduction to Polymer Physics

Theoretical Concepts and Applications

 Springer

Reinhard Hentschke
School of Mathematics and Natural
Sciences
Bergische Universität Wuppertal
Wuppertal, Nordrhein-Westfalen, Germany

ISSN 2192-4791 ISSN 2192-4805 (electronic)
Undergraduate Lecture Notes in Physics
ISBN 978-3-031-87323-2 ISBN 978-3-031-87324-9 (eBook)
https://doi.org/10.1007/978-3-031-87324-9

Preface

This text is intended for students, predominantly in physics and with a one-semester background in statistical mechanics, who are about to start their bachelor thesis project or who are past this stage already. These students may be interested in polymer physics, because they pursue research projects overlapping with this area of science or because they want to broaden their knowledge in condensed matter physics in general. If they have followed the regular physics curriculum up to this point, it is unlikely that they had much exposure to polymer physics. Hence they may be interested in an overview, without becoming overwhelmed, providing a basis from which to explore more detailed or advanced textbooks, research papers, and talks.

Why should you be interested in polymer physics? One reason is that polymer materials are ubiquitous and have countless applications. Hence, a significant amount of research and development goes into polymer-related products, which in turn means that there are well-paid jobs and not only for chemists! Many physics students don't know this, because they believe that polymers are an exclusive domain for chemists. Another reason why polymer physics is interesting is that problems in polymer physics require a broad background in physics, including both classical and quantum physics. There are virtually countless intellectually challenging questions and answering them never gets boring. Polymer physics overlaps with the physics of liquids, colloids and with the physics of surfaces and interfaces. Note that polymer materials are complex and may contain many components which are not polymers—like nanoparticles. There is of course overlap with chemistry in general and physical chemistry in particular. Can one make big discoveries in polymer physics? Most physics students, at least initially, have this dream. Well, Nobel prizes have been awarded for work in polymer science. However, since there is not just 'one big mystery' (like what is Dark Matter), progress and discoveries are achieved in smaller increments. But this usually means that earnest intelligent work will be rewarded with success and recognition—which cannot be said about the work on 'big questions'.

Virtually all the material compiled in these notes was worked out long ago and thus can be found in numerous textbooks. However, all textbooks, and these notes are no exception, present a subset of the 'complete polymer knowledge'. Sometimes the subset is larger and sometimes it is more focussed ('less' sometimes is indeed 'more'!). But always there is a certain bias due to the author's feel at the time of writing of what a student should know. A serious difficulty when teaching polymer physics are the many variables, e.g., polymer type and mass, environments like melts, solvents (good, bad, θ-), gels, networks, etc., temperature, frequency and so on and so forth. This gives rise to a host of results for numerous combinations of system parameters, mostly power laws, which can be very confusing for the beginner since no Ariadne thread appears to exist in this maze. Clearly, the challenge is to avoid this confusion without omitting too much important information.

The current standard textbook on polymer physics is M. Rubinstein and R. H. Colby *Polymer Physics*, Oxford University Press. This book focusses on the theoretical concepts, which does not mean that there are no experimental results to which the theory is compared. The opposite is true. However, these well defined experiments are designed specifically to test abstract theory. In practice, polymer materials are designed for complex applications requiring likewise complex materials. This in turn means that experimental measurements are less 'clear' and usually less 'clean'. Understanding an application polymer material on the molecular level is a very demanding task! It is therefore interesting to contrast Rubinstein and Colby with another book written by an application scientist. C. Wrana's *Introduction to Polymer Physics*, published by the Lanxess company (unfortunately this English version is difficult to obtain; widely available, however, is the German edition *Polymerphysik* published by Springer), focusses on the practical side—albeit for elastomers. Elastomers, the polymer ingredients of rubber materials, are arguably the 'most physical' of all polymers. Wrana concentrates on the viscoelastic properties of elastomer materials and their nonlinear deformation behaviour from the point of view of someone working in the elastomer industry. He also includes a discussion of nano-fillers and their effects on the mechanics of elastomers. These notes are an attempt to cover and combine key sections in both texts—nevertheless, bits and pieces from many other literature sources are included as well (and referenced throughout the notes). Here I briefly want to mention a mere few. Two older general textbooks on polymer physics are G. Strobl *The Physics of Polymers*, Springer, and R. J. Young and P. A. Lovell *Introduction to Polymers*, Chapman & Hall. Strobl discusses in particular many of the basic key measurements. The book by Young and Lovell is less mathematical and focusses more on synthesis and polymer characterization as well as on the description of numerous standard techniques used for characterization. And then there is one book that every person seriously working in polymer physics had and probably still has at hand—P.-G. de Gennes *Scaling Concepts in Polymer Physics*, Cornell University Press. The book was first published in 1979, but its elegant style and timeless ingenuity makes it a valuable source of insight even to this day. However, I do not recommend it for beginners. Pretty much the same comment, minus the elegance perhaps, applies to M. Doi and S. F. Edwards *The Theory of Polymer Dynamics*, Oxford Science Publications. However, this book contains most, if not all, the details, which makes it very valuable to the advanced student of the subject. There is a less formidable text by Doi alone (M. Doi *Introduction to Polymer Physics*, Oxford Science Publications), which, even though it still is not a text for beginners, is much easier to read. Written at about the same time as Doi's book, *Statistical Physics of Macromolecules*, AIP Press, by A. Y. Grosberg and A. R. Khokhlov also is not a beginner's text. Again, it is of interest for polymer students at a more advanced stage (e.g., its extensive discussion of phase transitions in polymer systems). Finally, going still back in time to the 1950s and 1960s, there are the books by Paul Flory *Principles of Polymer Chemistry*, Cornell University Press and *Statistical Mechanics of Chain Molecules*, Interscience Publishers. Reading these notes you will recognize his wide-ranging impact on the field. Like the previous books, I do not recommend them for beginners. But Flory explains many important aspects of theoretical approaches in polymer physical chemistry, which the subsequent literature does not explain with the same amount of detail and insight. You may want to keep this in mind and when you seem to be missing a piece of information, hampering your understanding of a problem in polymer physical chemistry, you should consult Flory's books. Both Flory (1974) and de Gennes (1991), by the way, have received Nobel prizes (Flory in Chemistry and de Gennes in Physics) for their work on the physical properties of polymers.

The computer simulation of polymers has produced and will continue to contribute important insights. However, it is difficult to develop this aspect of polymer research in a text like the present one including a sensible amount of detail. Therefore, no such attempt has been made. Nevertheless, I want to mention some references. *Computer Simulation of Liquids* by M. P. Allen and D. J. Tildesley (Oxford University Press: Oxford (1990)), in my opinion, still ranks among the best books on the subject of computer simulation for beginners. If somebody wants to try out some basic simulations fast, then I can offer my *Computer Simulation Laboratory* (https://constanze.materials.uni-wuppertal.de/fileadmin/physik/the ochemphysik/Skripten/SimTutorial.pdf). Specifically, for polymer simulations, one can find a number of texts. One example is *Simulation Methods for Polymers* edited by M. Kotelyanskii and D. N. Theodorou (Marcel Dekker: New York (2004)). The advanced reader will benefit from *Monte Carlo Simulations in Statistical Physics* by D. P. Landau and K. Binder (Cambridge University Press: Cambridge (2000)). Especially the second author and his group have made numerous important contributions to the simulation of polymers, which is reflected in the book.

As we move along, it may be useful and even necessary to look up physical parameters of polymers. For this purpose, I recommend J. E. Mark (editor) *Physical Properties of Polymers Handbook*. (2nd edition) Springer: NewYork (2007).

Wuppertal, Germany   Reinhard Hentschke

Acknowledgements

A number of colleagues and students have supported my efforts to present this material in useful fashion—either by providing general encouragement or by improving specific points or by pointing out mistakes. These people, to whom I am very grateful, are Nils Hojdis, Frack Fleck, Michael Karbach, Jan Weilert, and Lena Tarrach. In addition, I am grateful for being permitted to reprint or otherwise use figures and data from various publications mentioned in the text. Here I owe special thanks to Andrej Lang, who supplied me with high-quality versions of Figs. 4.14 and 4.16, as well as Claus Wrana, Jan Weilert, and Gert Heinrich for their permission to reprint figures from their works.

Wuppertal, Germany

Reinhard Hentschke

Contents

Chapter 1
Polymers—Microstructure, Classification, and Mass

Abstract First, we shall familiarize ourselves with the microstructure and some of the related classifications of polymers. In addition, we shall discuss a key variable—the mass of polymers and how different experiments measure different types of mass.

1.1 Polymer Microstructure and Classification

The simplest type of polymer is a linear covalent chain containing one type of monomer, i.e. -A-A-A-A-A-A-A-A-. This is a **homopolymer**. A few examples are compiled in Table 1.1. It is not obvious from -A-A-A-A-A-A-A-A- that there are possible differences between this -A-A-A-A-A-A-A-A- and another -A-A-A-A-A-A-A-A- due to stereoisomerism or structural isomerism. The former is depicted in Fig. 1.1. An example of structural isomerism is shown in Fig. 1.2. Since rotation around the double bond is not possible, polyisoprene, for instance, can locally exist as *cis* or *trans*. How much *cis* or *trans* there is depends on how the polymer was polymerized. Naturally produced PI from rubber trees (hevea brasiliensis) is almost completely cis-1,4-polyisoprene. Even though these stereoisomers or structural isomers do not 'look' very different, their physical properties are usually quite distinct. For instance, isotactic homopolymers show the tendency to form helical structures, whereas their atactic variants do not. Another example is cis-1,4-polyisoprene, which is used in large quantities in automobile and particularly in truck tires, whereas trans-1,4-polyisoprene (gutta-percha) has a strong tendency to crystallize and today has a few specialized uses only.

The combination of several different monomers, as shown in Fig. 1.3, yields so called **heteropolymers**. If the heteropolymer contains two monomer types it is called a **copolymer**. Copolymers can be alternating, random, block or graft. Polymers containing two blocks are called diblock copolymers; with three blocks they become triblock copolymers and then multiblock copolymers. Polymers containing three types of monomers are called **terpolymers**. An example of a biopolymer with four different types of monomers (nucleotides) is DNA.

Another distinguishing feature is the polymer architecture. Figure 1.4 shows sketches of (a) linear, (b) ring, (c) star-branched, (d) ladder, (e) comb, (f) dendrimer,

R. Hentschke, *A Concise Introduction to Polymer Physics*, Undergraduate Lecture Notes in Physics, https://doi.org/10.1007/978-3-031-87324-9_1

Table 1.1   A few common homopolymers

Monomers	Polymer	Comments
Ethylene $CH_2 = CH_2$	Polyethylene (PE) $-[CH_2 - CH_2]_n-$	Moulded objects, tubing, films, insulation
Propylene $CH_2 = C(CH_3)H$	Polypropylene (PP) $-[CH_2 - C(CH_3)H]_n-$	Similar uses to PE
Styrene $CH_2 = C(C_6H_5)H$	Polystyrene (PS) $-[CH_2 - C(C_6H_5)H]_n-$	Cheap moulded objects
Vinyl chloride $CH_2 = C(Cl)H$	Poly(vinyl chloride) (PVC) $-[CH_2 - C(Cl)H_2]_n-$	Pipes, hoses, sheathing on electrical cables
Isoprene $CH_2 = C(CH_3)-CH=CH_2$	cis-1,4-polyisoprene (PI) $-[CH_2 - C(CH_3)=CH-CH_2]_n-$	Tires, latex products Sheathing on electrical cables

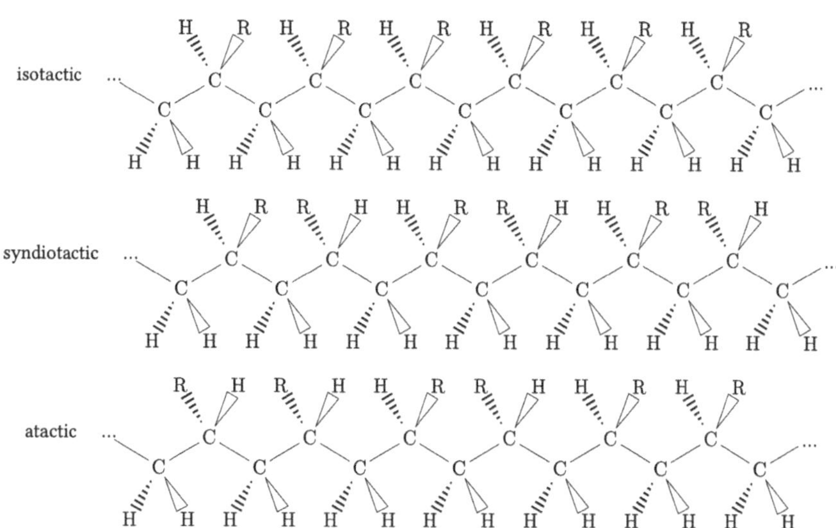

Fig. 1.1   Tacticities of vinyl polymers. The carbon atoms are in the same plane, whereas the hydrogens and the R-moieties lie below or above, respectively

Fig. 1.2   Two structural isomers of polyisoprene

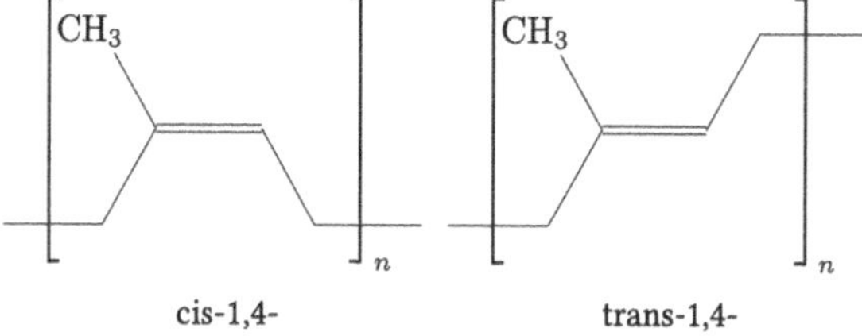

$$\cdots-A-B-A-B-A-B-A-B-A-B-\cdots$$
alternating

$$\cdots-A-B-B-B-A-A-B-A-B-A-\cdots$$
random

$$\cdots-A-A-A-A-A-B-B-B-B-B-\cdots$$
diblock

$$\cdots-A-A-A-\cdots-B-B-B-\cdots-A-A-A-\cdots$$
triblock

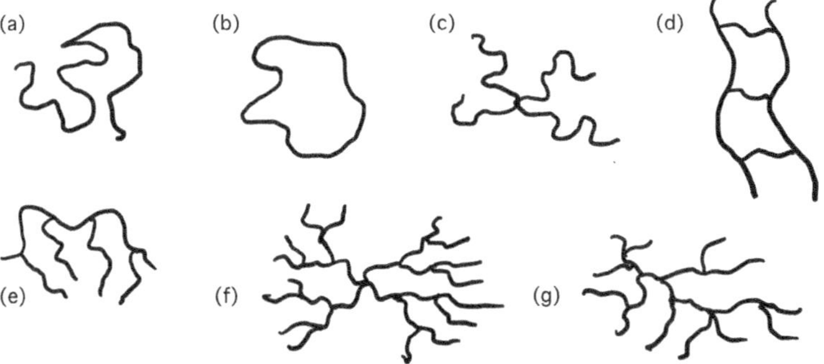

graft

$$\cdots-A-A-B-B-\cdots-B-B-A-A-\cdots-A-A-B-B-\cdots-B-B-A-A-\cdots$$
multiblock

$$\cdots-A-B-B-C-C-B-A-B-A-B-B-B-A-C-C-B-\cdots$$
random

$$\cdots-A-A-A-\cdots-B-B-B-\cdots-B-B-B-\cdots-C-C-C-\cdots$$
ABC-triblock

Fig. 1.3 Different types of polymers. Figure adopted from Rubinstein and Colby

Fig. 1.4 Examples of polymer architectures: **a** linear, **b** ring, **c** star-branched, **d** ladder, **e** comb, **f** dendrimer, and **g** randomly branched

and (g) randomly branched polymer architectures. In this text our sole focus is case (a), linear polymers.

Technical polymers are usually divided into three groups, as shown in Fig. 1.5:

Thermoplastics are linear or branched polymers that can melt when heated. They can be moulded and remoulded using processing techniques like injection moulding and extrusion. They constitute the largest fraction of industrial polymers. Many thermoplastics are amorphous and are incapable of crystallization. Amorphous polymers are characterized by their **glass transition temperature** T_g, the temperature at which they transform from a hard or glassy state to a soft or rubbery state.

If the polymers crystallize, this occurs between the glass transition, T_g and the melting temperature T_m. In this temperature range, there is sufficient molecular movement for crystalline domains to form. This process can take place during heating or during cooling from the melt. The main characteristics of crystalline polymers is that they are normally semi-crystalline. The size, shape, and percentage of crystals

Fig. 1.5 Polymer
classification

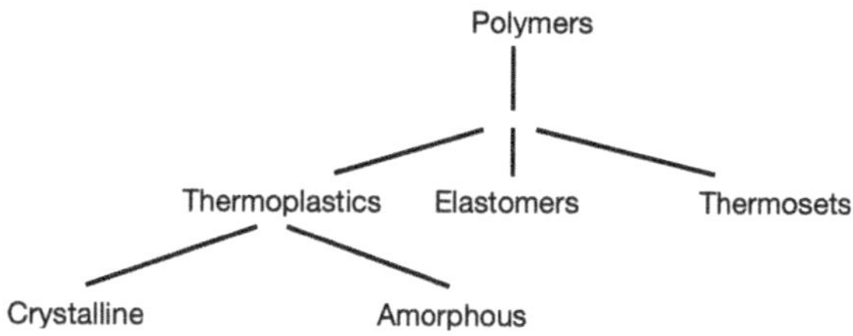

depend on the heating and cooling of the material. The change from the amorphous-liquid state of the melt to the crystalline state is a kinetically controlled process and depends primarily on nucleation. This is why the crystallization temperature T_c is always below the thermodynamically controlled melting temperature (for a detailed discussion see Chap. 4 in [1]).

Elastomers are polymers whose response to large deformations is mostly elastic over a wide range of temperatures. This range is bracketed by T_g on the low side and the thermal degradation on the high side. Elastic response to deformations is greatly enhanced by cross-linking of the polymers, which largely prevents them from 'flowing' in response to the deformation. Such materials are commonly known as rubbers.

Since we just mentioned 'cross-linking'. What is a **cross-link**? A **chemical cross-link** consists of a certain number -usually in the single digits- of covalent bonds joining two polymer chains or, since polymers are flexible and long, form a link within one and the same polymer chain. A famous example is the formation of sulfur (atom) bridges in a reaction called **vulcanization**. The second type of cross-link is the **physical cross-link**. Polymers are mostly like 'entangled ropes'. These entanglements are hairpin-style curves of one polymer around another or one polymer section around another section of the same polymer. The density and the nature of the cross-links is one of the most important parameters determining the physical behavior of polymer systems.

Thermosets are generally rigid materials and highly cross-linked polymers. Like elastomers they are intractable once formed and degrade rather than melt when heated.

All that said, it should be borne in mind that this classification scheme is not absolute and that there is some overlap between the three groups.

When we look at polymers from the perspective of statistical thermodynamics, they possess one particular feature—**conformation entropy**, distinguishing them from simple liquids or solids. A chain of monomers, even a short one between cross-links or other branch or terminal points, can have many different shapes. The number of possible shapes or 'paths' give rise to the conformation entropy of the chain. This entropy wants to be a large as possible. A constraint imposed on the number of possible chain conformations reduces the conformation entropy and produces forces acting along the entire chain and beyond. This long range effect in conjunction with liquid-like interactions produces unique structural and dynamic material properties.

1.2 Molecular Mass Distribution

When discussing polymer microstructure, we noted that two chemically identical polymers can be very different if they are isomers. Another distinguishing feature is mass. Nature can produce biopolymers possessing a specific mass. Technical polymers, on the other hand, are more or less **polydisperse**, i.e. they possess a certain mass distribution. The **number-average molar mass** $\bar{M}_n$ is defined via

$$\bar{M}_n = \sum_i x_i M_i \; . \tag{1.1}$$

The index i indicates the fraction of polymers in the sample containing N_i molecules of molar mass M_i. The quantity x_i is the mole fraction of these polymers, i.e.

$$x_i = N_i / \sum_j N_j \; . \tag{1.2}$$

A related quantity is the **weight-average molar mass** $\bar{M}_w$ defined via

$$\bar{M}_w = \sum_i w_i M_i \; , \tag{1.3}$$

where

$$w_i = N_i M_i / \sum_j N_j M_j \; . \tag{1.4}$$

There are still other average masses, but just $\bar{M}_n$ and $\bar{M}_w$ alone may be confusing. What is the difference between them?

Note that x_i is the probability to pick any polymer from group i at random, whereas w_i is the probability to pick any monomer from within any polymer in this group. These probabilities can be very different. Imagine 3 identical light polymers and one heavy polymer, all composed of the same type monomer. If $i = 1$ stands for the light polymer, then $x_1 = 3/4$. If we now cut the light polymers in half and add the monomers which we have cut off to the heavy polymer—what changes? Well, x_1 is still 3/4, but w_1 is only half of what it was before (Note: Even though x_i does not change, $\bar{M}_n$ changes during this 'experiment'.). The only time $x_i = w_i$ is when there is only one group or one term in $\sum_i$, i.e. all polymers possess the same mass, which means that they are **monodisperse**. In this case $\bar{M}_w / \bar{M}_n = 1.0$. Hence, we introduce x_i and w_i as a means to determine polydispersity. We can see this more clearly via the following equation:

$$\bar{M}_w = \bar{M}_n + \frac{\delta \bar{M}_n^{\,2}}{\bar{M}_n} \; . \tag{1.5}$$

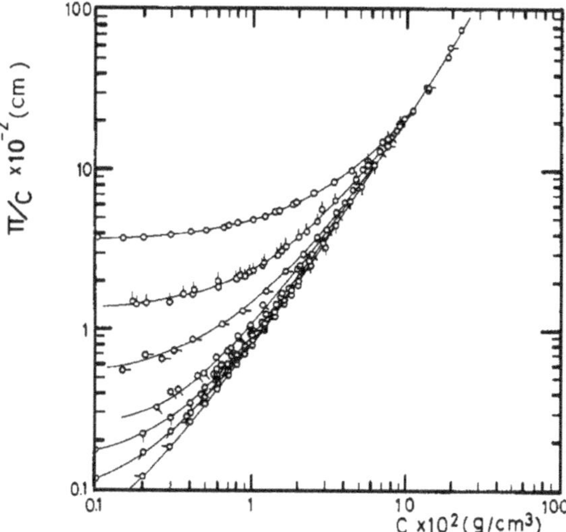

Fig. 1.6 Concentration dependence of the osmotic pressure of poly(α-methyl styrene) for different molecular weights in toluene at $T = 25\,^\circ$C. Reprinted with permission from [2]. Copyright 1981 American Chemical Society

Here $\delta \bar{M}_n^{\,2} = \bar{M}_n^{\,2} - (\bar{M}_n)^2$ is the variance of the mass computed with the distribution (1.2) (see Problem 1.1). Typically $\bar{M}_w/\bar{M}_n$ is in the range 1.5–2.0. But how do we measure these different molar masses?

The **osmotic pressure** Π of polymers in solution in the limit of vanishing polymer concentration c (usually in g/volume) is given by **van't Hoff's law**:

$$\frac{\Pi}{c} = \frac{RT}{\bar{M}_n} \, . \tag{1.6}$$

A formula for Π in polymer solutions will be discussed in Chap. 3. However, we can already see that Π involves $\bar{M}_n$ rather than $\bar{M}_w$. Had we performed the above 'thought experiment', which does not alter the number of polymers, in the polymer solution, it would not have altered Π, since the osmotic pressure in van't Hoff's law depends on the polymer concentration only and not on the distribution of monomers throughout the polymer population. Figure 1.6 shows the concentration dependence of the osmotic pressure of poly(α-methyl styrene) for different molecular weights in toluene. Note that (1.6) indeed describes the limit of vanishing concentration only. But this is sufficient to extract $\bar{M}_n$ (see Problem 1.2).

Remark: We shall return to Fig. 1.6 and the apparent $\bar{M}_n$-independence of Π/c at high concentrations in Problem 3.5, when we have discussed all theoretical concepts necessary to understand this observation.

And how do we measure $\bar{M}_w$? One method is static light scattering. In order to understand this, we must invest some work. Nevertheless, the effort is worth it, because it exemplifies the sensitivity of different experiments to the various types of average mass.

Let us assume that unpolarized light, possessing the wavelength λ, interacts with an isolated isotropically polarizable molecule. The linear dimension of the molecules is also assumed to be much less than λ. Within the molecule a dipole moment $\vec{p} = \alpha \vec{E}(\omega)$ is induced by the local $\vec{E}$-field. Here α is the isotropic **polarizability** of the molecules, $\omega = 2\pi c/\lambda$, and c is the speed of light. Now remember the discussion about dipole radiation from your E&M lecture. The pointing vector $\vec{S}$, measuring electromagnetic energy per area and time in a certain direction, which essentially is our scattered intensity, is given by

$$\vec{S} = \frac{1}{4\pi r^2 c^3}(\ddot{\vec{p}} \times \vec{n})^2 \vec{n} \tag{1.7}$$

in the case of dipole radiation. Here r is the distance between the dipole and the detector and $\vec{n}$ is a unit vector in the direction from the dipole to the detector. The double dot indicates time derivatives. We also remember that the magnitude of the Pointing vector (at the position of the molecule) is $S_{in} = cE^2/(4\pi)$. Now we can write down the **Rayleigh ratio** R_θ, which is the scattered intensity times r^2 divided by the incident intensity, i.e.

$$R_\theta = \frac{16\pi^4}{\lambda^4}\alpha^2 \rho P(\theta) \,, \tag{1.8}$$

where

$$P(\theta) = \langle(\vec{e}_p \times \vec{n})^2\rangle_{\vec{e}_p} \,. \tag{1.9}$$

The quantity $\vec{e}_p$ is a unit vector in $\vec{p}$-direction. In addition, $\langle\ldots\rangle_{\vec{e}_p}$ is an orientation average of the dipole vector and we have used $\ddot{E} = \omega^2 E$. The factor ρ is an average number density of isolated molecules, which we insert here since we want to collect intensity from $\langle N\rangle$ molecules per volume V, i.e. $\rho = \langle N\rangle/V$. The quantity $\langle\ldots\rangle$ is a thermodynamic average of the (in principle) fluctuating number of molecules.

If we define the direction of the incident light as our z-direction, then $\vec{e}_p = (\cos\varphi, \sin\varphi, 0)$, i.e. $\vec{e}_p$ is in the xy-plane. Note that the components of the $\vec{E}$-field, and therefore the components of the induced dipole, are perpendicular to the direction in which the light travels. In the same coordinate system $\vec{n} = (\cos\phi\sin\theta, \sin\phi\sin\theta, \cos\theta)$. With this we find

$$P(\theta) = (\cos^2\theta + 1)/2 \,. \tag{1.10}$$

However, we are not so much interested in isolated molecules. Instead we are interested in a dilute solution of our polymers in a solvent. If ϵ_o is the dielectric constant of the pure solvent and ϵ_1 is the dielectric constant of the solution, then we have $(\epsilon_o - 1)/(4\pi)\vec{E} = \vec{P}_o$ and $(\epsilon_1 - 1)/(4\pi)\vec{E} = \vec{P}_1$, where $\vec{E}$ is the average electric field and the polarization $\vec{P}$ is the local dipole moment per unit volume. Subtracting the two equations yields $(4\pi)^{-1}(\epsilon_1 - \epsilon_o)\vec{E} = \Delta\vec{P}$. Note that $V\Delta\vec{P}$

replaces or previous $\vec{p}$ when the molecules were isolated (cf. Debye's article from 1947 cited below!). We now write

$$\frac{\epsilon_1 - \epsilon_o}{4\pi} = \frac{n_1^2 - n_o^2}{4\pi} \approx \frac{1}{4\pi} 2n_o \frac{dn}{dc_w}\bigg|_o c_w \ . \tag{1.11}$$

The quantities n_1 and n_o are the refractive indices of the solution and the solvent, respectively. Note that we expand $n_1(c_w)^2$, where c_w is the solute weight concentration, in terms of c_w around n_o. With this the Rayleigh ratio becomes

$$R_\theta = \left(\frac{2\pi}{\lambda}\right)^4 \frac{1}{(4\pi)^2} \frac{1}{V} \left(2n_o \frac{dn}{dc_w}\bigg|_o\right)^2 c_w^2 V^2 P(\theta) \ . \tag{1.12}$$

Aside from the above replacement of the molecular dipole moment $\vec{p}$ by $V \Delta P$ we have also replaced ρ by $1/V$. The average number of molecules $\langle N \rangle$ now becomes the average number of polymers in the system $\langle \nu_1 \rangle$ and, as we shall see, $\langle \nu_1 \rangle$ is contributed by a new version of $P(\theta)$.

The overall solute mass density c_w can be rewritten in terms of an integral over the local solute mass densities $c_w(\vec{r})$ throughout the system, i.e.

$$c_w = \frac{1}{V} \int_V d^3r \, c_w(\vec{r}) \ . \tag{1.13}$$

Thus far $P(\theta)$ only accounts for radiation without interference of radiation (amplitudes) from the different volume elements in the system (you can find a detailed discussion of the steps surrounding '$\rightarrow$' in the following formula in Chap. 9 of [3]; an excellent book worth owning even in the age of the internet!). This means we need another substitution, i.e.

$$c_w^2 P(\theta) \rightarrow \langle c_w \rangle^2 P_i(\theta) \equiv \frac{\langle c_w \rangle^2}{V^2} \int_V d^3r \, d^3r' e^{i\vec{q}\cdot(\vec{r}-\vec{r}')} \frac{\langle c_w(\vec{r})c_w(\vec{r}')\rangle}{\langle c_w \rangle^2} \ . \tag{1.14}$$

Here $\vec{q} = \vec{k}_{out} - \vec{k}_{in}$ is the momentum transfer from the incoming wave vector $\vec{k}_{in}$ to the scattered one $\vec{k}_{out}$, i.e. we keep the phases of $\vec{E}_{in}(\vec{r})$ and $\vec{E}_{out}(\vec{r})$ and let them interfere. Note that the index i is not a running index. It is meant to distinguish $P_i(\theta)$ from $P(\theta)$. The above corresponds to the 1st Born approximation in quantum scattering theory.

Note also that we collect scattering intensity from many thermodynamically identical copies of our system of interest. Hence our final R_θ is a thermal average over such systems, i.e.

$$R_\theta = K\langle c_w \rangle M P_i(\theta)\langle \nu_1 \rangle \ , \tag{1.15}$$

where M is the polymer mass, i.e. $\langle c_w \rangle = M\langle \nu_1 \rangle / V$, and

$$K = \frac{4\pi^2 n_o^2}{\lambda^4} \left(\frac{dn}{dc_w} \bigg|_o \right)^2 . \tag{1.16}$$

These expressions for R_θ and K date back to work by Einstein [4] and Debye [5]. The derivation of K is not trivial—the above is a mere sketch!—and initially there was even controversy regarding the exact form of the factor $n_o^2 \left(\frac{dn}{dc_w} \big|_o \right)^2$ (see [6]; first column on page 160).

Thus far we have considered monodisperse molecules. What happens if the molecules are polydisperse? In this case $c_w M$ is replaced by $\sum_i c_{w,i} M_i$, i.e.

$$c_w M \rightarrow \sum_i c_{w,i} M_i = \frac{\sum_i c_{w,i} M_i}{\sum c_{w,i}} c_w = \bar{M}_w c_w , \tag{1.17}$$

where $c_{w,i} = N_i M_i / V$ and $c_w = \sum_i c_{w,i}$. Hence,

$$R_\theta = K \langle c_w \rangle \bar{M}_w P_i(\theta) \langle v_1 \rangle \tag{1.18}$$

relates the light scattering intensity to the weight-average (molar) mass (there is no mass dependence left in $P_i(\theta)$). We shall come back to scattering and to (1.18) in Chap. 2. In particular we shall discuss the evaluation of $P_i(\theta)$ in detail.

Of course there are more techniques for measuring molar masses, which we must bypass at this point. An overview can be found in [7]. A nice review covering light scattering is [8].

Remark: If you want information regarding a particular technique, e.g., light scattering, it is often a good idea to check out the webpages of the companies selling this particular experimental equipment. Typically you can find 'technical glossaries', 'learning pages', etc. containing very good, albeit simplified, explanations of the method and the underlying theory.

We shall also skip the shape of the mass distribution, since it depends on the polymerization mechanism—a topic we do not cover here.

1.3 Problems

Problem 1.1 Show that (1.5) is correct.

Problem 1.2 Use Fig. 1.6 to estimate $\bar{M}_n$ for the top three curves. Compare your numbers to the numbers in Table I of [2].

References

1. R.J. Young, P.A. Lovell, *Introduction to Polymers*, 2nd edn. (Chapman & Hall, London, 1991)
2. I. Noda, N. Kato, T. Kitano, M. Nagasawa, Thermodynamic properties of moderately concentrated solutions of linear polymers. Macromolecules **16**, 668 (1981)
3. J.D. Jackson, *Classical Electrodynamics* (Wiley, New York, 1975)
4. A. Einstein, Theorie der Opalenszenz homogener Flüssigkeitsgemische in der Nähe des kritischen Zustandes. Annalen der Physik **33**, 1275 (1910)
5. P. Debye, Molecular weight determination by light scattering. J. Phys. Chem. **51**, 18 (1947)
6. P.M. Doty, B.H. Zimm, H. Mark, An investigation of the determination of molecular weights of high polymers by light scattering. J. Chem. Phys. **13**, 159 (1945)
7. S.E. Harding, K. Varum, B. Stoke, O. Smidsrod, Molecular weight determination of polysaccharides. Adv. Carbohydr. Anal. **1**, 63 (1991)
8. P.J. Wyatt, Light scattering and the absolute characterization of macromolecules. Anal. Chimica Acta **272**, 1 (1993)

Chapter 2
Equilibrium Conformation of Single Chains

Abstract Many of the bulk properties of polymers are intimately related to the flexibility of the polymer chains or the flexibility of chain segments as part of a complex polymer architecture. In this section, we investigate the statistical thermodynamics of individual chains by developing and applying various theoretical concepts. In addition, we discuss scattering and how it can be used to validate our theoretical models.

2.1 Flexibility Mechanism and Polymer Dimension

A quantity used to characterise the linear dimension of a polymer chain is its **end-to-end vector** $\vec{R}$, which is the sum over monomer end-to-end vectors $\vec{b}$ (cf. Fig. 2.1). The monomer end-to-end vector may itself be a sum over a number of shorter vectors within the flexible monomer as shown in the example in Fig. 2.2. Having said this, we want to concentrate on the simplest case, i.e. in the following the mathematical realization of a polymer chain is the sum over bond vectors $\vec{b}_i$ of constant magnitude b as depicted in Fig. 2.3.

But what do we mean by **flexibility** in this context? Figure 2.4 depicts four covalently bonded atoms i, j, k, and l. The bond lengths are pretty much constant and so are the valence angles, i.e. the angles between the bonds ij and jk or jk and kl. But when the bond jk is a single bond, then the angle ϑ, which is called **torsion angle**, is variable. A polymer contains a large number of such torsion angles, which allow the polymer chain to assume many different **conformations**. Loosely speaking, flexibility is a measure for the diminishing orientation correlation between two vectors $\vec{b}_i$ and $\vec{b}_{i+s}$ along a polymer chain as function of their separation s. In other words, rotational freedom of the torsion angles reduces orientational correlation between $\vec{b}$-vectors and thus increases chain flexibility.

Under these conditions, i.e. fixed bond lengths and fixed valence angles while the torsion angles can vary freely, what is the largest end-to-end distance R_{max}? Let's look at Fig. 2.3 and assume that all bond lengths b_i and valence angles ϕ_i are constant, i.e. $b_i = b$ and $\phi_i = \phi \; \forall \, i$. If we flatten our polymer onto a planar surface,

© The Author(s), under exclusive license to Springer Nature Switzerland AG 2025

R. Hentschke, *A Concise Introduction to Polymer Physics*, Undergraduate Lecture Notes in Physics, https://doi.org/10.1007/978-3-031-87324-9_2

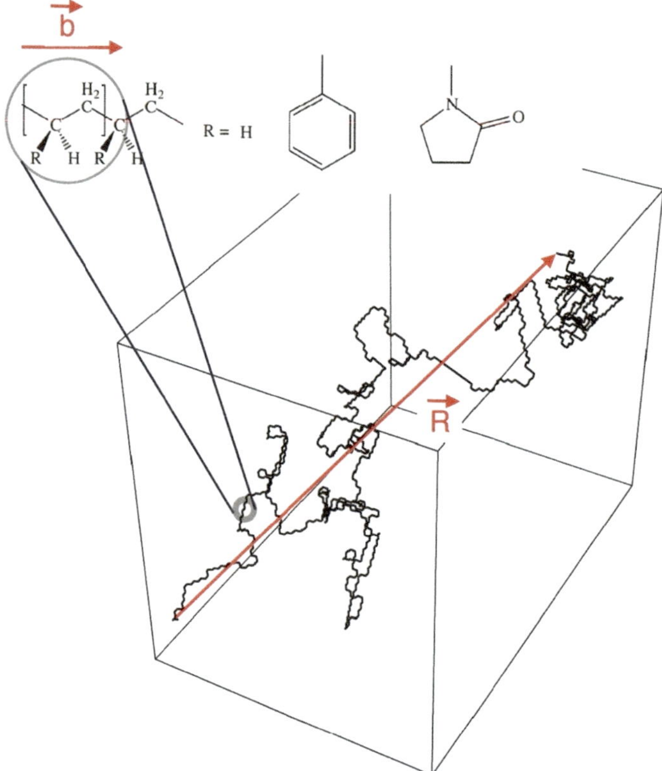

Fig. 2.1 Examples (cartoon!) of a linear homopolymer chain based on the monomer unit -CH$_2$-CHR, where R stands for different moieties (three examples are shown: polyethylene, polystyrene and polyvinylpyrrolidone). Note that $\vec{R}$ (not to be confused with the previous letter R) is the end-to-end vector, which itself is a sum over monomer vectors $\vec{b}$

Fig. 2.2 Monomer unit of p-PHB (poly(4-hydroxybenzoic acid)). The ϕ_i are valence angles, whereas the ϑ_i are torsion angles

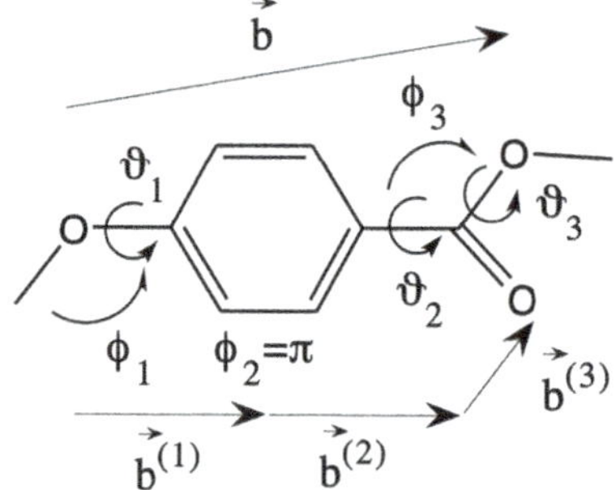

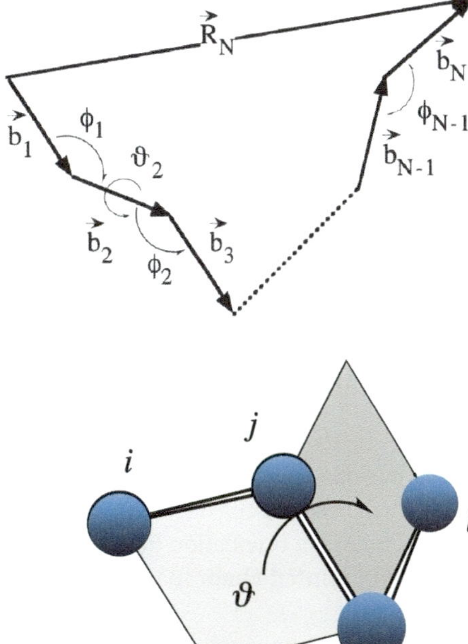

Fig. 2.3 Cartoon of a linear polymer consisting of N units

Fig. 2.4 Definition of a torsion angle via four successive atoms

we obtain the largest end-to-end distance for a straight zig-zag line. The projection of each bond onto the direction defined by this zig-zag line is $b \cos((\pi - \phi)/2)$ and therefore the overall length of the zig-zag line is

$$R_{max} = b \cos((\pi - \phi)/2)N = b \sin(\phi/2)N. \tag{2.1}$$

This length, i.e. the maximum magnitude of the end-to-end vector, is also called the **contour length** L of the polymer, i.e. $L = R_{max}$.

It is important to note that the definition of the contour length depends on the flexibility mechanism. In the present case this means that measuring contour length along the bonds of the polymer backbone yields a greater length, i.e. bN.

While a polymer only has one contour length, it can assume many different conformations and consequently the magnitude (and direction) of its end-to-end vector varies. Therefore we focus on statistical averages of the powers of $\vec{R}$. Obviously

$$\langle \vec{R} \rangle = 0, \tag{2.2}$$

i.e. this average does not tell us much. However, $\langle \vec{R}^2 \rangle \neq 0$ and since $\sqrt{\langle \vec{R}^2 \rangle}$ has the unit of length we are going to use it as measure for the linear dimension of a polymer in space.

So let's study the end-to-end vector depicted in Fig. 2.3 in more detail, i.e.

$$\vec{R}_N = \sum_{i=1}^{N} \vec{b}_i = b \sum_{i=1}^{N} \vec{e}_i. \tag{2.3}$$

Note that the $\vec{e}_i$ are unit vectors. The quantity we are most interested in is the average of the square of $\vec{R}_N$:

$$\langle \vec{R}_N^2 \rangle = b^2 \sum_{i=1}^{N} \sum_{j=1}^{N} \langle \vec{e}_i \cdot \vec{e}_j \rangle = b^2 \left(1 + \frac{2}{N} \sum_{i>j}^{N} \langle \vec{e}_i \cdot \vec{e}_j \rangle \right) N. \tag{2.4}$$

Generally, the correlation function $\langle \vec{e}_i \cdot \vec{e}_j \rangle$ is quite difficult to calculate. It depends not only on $i - j$ but on i and j individually, since the middle of a polymer is different from its ends.

Let us therefore consider two simple approximations. The first one is $\langle \vec{e}_i \cdot \vec{e}_j \rangle = 0$ ($i \neq j$), i.e. the orientation of $\vec{e}_i$ does not influence $\vec{e}_j$ in the slightest. This defines the **freely jointed chain** for which

$$\langle \vec{R}_N^2 \rangle = b^2 N. \tag{2.5}$$

This result has the familiar form of the **Einstein law of diffusion**

$$\langle \vec{R}(t)^2 \rangle = 6Dt. \tag{2.6}$$

Here $\vec{R}(t)$ is the vector connecting the starting and the end point of a **random walk** in three dimensions (3D) measured at time t after the walker has started and D is the (long time) **diffusion coefficient**. Note that the time t is the equivalent of N. In other words, a polymer chain is just a random walk.

The second approximation is the straight zig-zag chain which we had used to calculate R_{max}. Here $\langle \vec{e}_i \cdot \vec{e}_j \rangle = 1$ if $i - j$ is even and $\langle \vec{e}_i \cdot \vec{e}_j \rangle = \cos(\pi - \phi)$ if $i - j$ is odd. Hence, $\sum_{i>j}^{N} \langle \vec{e}_i \cdot \vec{e}_j \rangle = N(N-1)/2 \cdot (1 - \cos\phi)/2$ and thus

$$\langle \vec{R}_N^2 \rangle = b^2 \sin^2(\phi/2) N^2, \tag{2.7}$$

where we assume that N is large. Note that (2.7) is consistent with (2.1).

A generalization of (2.5) is

$$\langle \vec{R}_N^2 \rangle = C_N b^2 N^{2\nu}. \tag{2.8}$$

In (2.5) the coefficient C_N equals 1 and the exponent ν equals 1/2. This value of ν characterizes an **ideal chain**. If the approximation $\langle \vec{e}_i \cdot \vec{e}_j \rangle = 0$ does not hold, then the coefficient C_N will be larger than one. This is because non-vanishing correlations increase the chain's local stiffness and therefore the volume of the 'coil', i.e. $\langle \vec{R}_N^2 \rangle^{3/2}$

increases. In this context C_N is called **characteristic ratio**—even though in most cases this definition refers to the limit $N \rightarrow \infty$. However, since N does not have to be larger than $O(10)$ monomers in order for C_N to be close to C_∞ this is not important for most practical purposes. Whether or not the exponent ν also changes depends on the range of the correlation function $\langle \vec{e}_i \cdot \vec{e}_j \rangle$. We shall return to this point below.

The leading approximation of $\langle \vec{R}_N^2 \rangle$, which takes the angles $\phi_1, \vartheta_1, \phi_2, \vartheta_2, \ldots$ (cf. Fig. 2.3) into consideration, is the following:

$$\langle \vec{e}_i \cdot \vec{e}_j \rangle = \langle \begin{pmatrix} 1 \\ 0 \\ 0 \end{pmatrix} \Pi_{k=0}^{i-1-j} \mathbf{T}_{j+k \rightarrow j+k+1} \begin{pmatrix} 1 \\ 0 \\ 0 \end{pmatrix} \rangle$$

$$\approx \begin{pmatrix} 1 \\ 0 \\ 0 \end{pmatrix} \langle \mathbf{t} \rangle^{i-j} \begin{pmatrix} 1 \\ 0 \\ 0 \end{pmatrix}. \tag{2.9}$$

What does this mean? In the first equation each of the two unit vectors $\vec{e}_i$ and $\vec{e}_j$ are considered unit vectors along the respective x-axes of their own 'private' coordinate systems. The projection of $\vec{e}_j$ onto the direction of $\vec{e}_i$ is then handled in a series of $i - j$ steps or rotations of the coordinate system of $\vec{e}_j$ onto the coordinate system of $\vec{e}_i$. The average is over a matrix $\mathbf{T}$ that contains all angles ϕ and ϑ between monomers i and j. The second equation assumes that all steps are statistically independent, i.e. the average of a product of rotation matrices becomes the product of $i - j$ identical average single step rotation matrices $\mathbf{t}$. This is a strong approximation—but it simplifies matters considerably. Note that different monomers may be correlated despite this approximation. This is because a monomer's orientation is coupled to the orientation of its predecessor if for instance either the bond angle ϕ or the torsion angle ϑ (or both) is fixed or limited to a sector of the full solid angle. Nevertheless, for an 'observer' on the predecessor monomer, any average involving functions of ϕ and ϑ, defining the orientation of the following monomer, are independent of the position of the predecessor along the chain. This also implies that monomers in the middle of the chain are no different from those near the chain ends.

So how do we find $\mathbf{t}$? Figure 2.5 shows two successive vectors $\vec{b}_i$ and $\vec{b}_{i+1}$ along the x-axes of their coordinate systems. Let's assume we are looking at this from within the xyz-system and we want to rotate $\vec{b}_{i+1}$ so that it becomes parallel to the

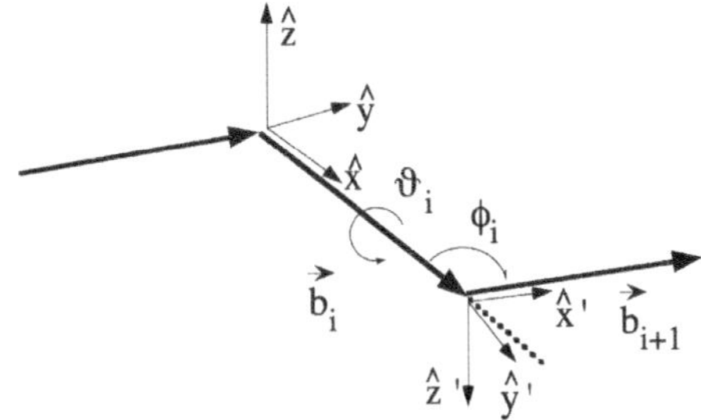

Fig. 2.5 Coordinate systems of two successive bonds

x-axis. First (1) we rotate $\vec{b}_{i+1}$ with respect to the x-axis by the angle $-\vartheta$ (omitting the index i) and then (2) we rotate the resulting $\vec{b}_{i+1}$ with respect to the z-axis until it is aligned with the x-axis. This angle of rotation is $\pi - \phi$ (also omitting the index i). Mathematically this is accomplished via the product of two rotation matrices:

$$\underbrace{\begin{pmatrix} -\cos\phi & \sin\phi & 0 \\ -\sin\phi & -\cos\phi & 0 \\ 0 & 0 & 1 \end{pmatrix}}_{(2)} \underbrace{\begin{pmatrix} 1 & 0 & 0 \\ 0 & \cos\vartheta & -\sin\vartheta \\ 0 & \sin\vartheta & \cos\vartheta \end{pmatrix}}_{(1)} \tag{2.10}$$

Note that this is not yet $\mathbf{t}$. Two things still need to be done. First, what we want is the inverse of the above matrix. In addition, we want to rotate coordinate systems rather than vectors within them. This means we must replace ϑ by $-\vartheta$ and ϕ by $-\phi$. The resulting $\mathbf{t}$ is given by

$$\mathbf{t} = \begin{pmatrix} -\cos\phi & \sin\phi & 0 \\ -\sin\phi\cos\vartheta & -\cos\phi\cos\vartheta & -\sin\vartheta \\ -\sin\phi\sin\vartheta & -\cos\phi\sin\vartheta & \cos\vartheta \end{pmatrix} . \tag{2.11}$$

Its average is

$$\langle\mathbf{t}\rangle = \begin{pmatrix} -\cos\phi & \sin\phi & 0 \\ -\sin\phi\langle\cos\vartheta\rangle & -\cos\phi\langle\cos\vartheta\rangle & -\langle\sin\vartheta\rangle \\ -\sin\phi\langle\sin\vartheta\rangle & -\cos\phi\langle\sin\vartheta\rangle & \langle\cos\vartheta\rangle \end{pmatrix} , \tag{2.12}$$

where we assume that ϕ is a constant angle.

Next we insert the approximation (2.9) into (2.4) and evaluate $\sum_{i>j}\langle\mathbf{t}\rangle^{i-j}$:

$$\sum_{i>j}\langle\mathbf{t}\rangle^{i-j} = (N-1)\langle\mathbf{t}\rangle + (N-2)\langle\mathbf{t}\rangle^2 + \cdots + (N-N)\langle\mathbf{t}\rangle^N$$

$$= N\sum_{i=1}^{N}\langle\mathbf{t}\rangle^i - \sum_{i=1}^{N} i\langle\mathbf{t}\rangle^i$$

$$= N\langle\mathbf{t}\rangle(\mathbf{I}-\langle\mathbf{t}\rangle)^{-1} - \langle\mathbf{t}\rangle(\mathbf{I}-\langle\mathbf{t}\rangle^N)(\mathbf{I}-\langle\mathbf{t}\rangle)^{-2}$$

$$\approx N\langle\mathbf{t}\rangle(\mathbf{I}-\langle\mathbf{t}\rangle)^{-1} \quad (\text{for large}N) \tag{2.13}$$

(cf. Problem 2.1). In the limit of large N our final result is

$$\langle\vec{R}_\infty^2\rangle = C_\infty b^2 N \tag{2.14}$$

with

$$C_\infty \frac{\mathbf{I}+\langle\mathbf{t}\rangle}{\mathbf{I}-\langle\mathbf{t}\rangle}\bigg|_{11}, \tag{2.15}$$

where the subscript 11 indicates the matrix element. When $\langle \sin \vartheta \rangle = 0$, i.e. the torsion potential is symmetric, this expression simplifies to

$$C_\infty = \frac{(1 - \cos \phi)(1 - \langle \cos \vartheta \rangle)}{(1 + \cos \phi)(1 + \langle \cos \vartheta \rangle)}. \tag{2.16}$$

Figure 2.6 shows a number of examples for C_N and C_∞ calculated at various levels of approximation. The dotted line in the top panel is the freely jointed chain for which, according to (2.5), $C_N = 1$. The dashed line in the same panel is for a freely rotating chain for which the angle between bonds is $\phi = 112°$ (corresponding to the C-C-C valence bond in polyethylene (PE)). Since $\langle \cos \varphi \rangle = 0$ in this case, (2.16) yields $C_\infty \approx 2.2$. However, the figure also shows that C_N increases with N. The N-dependence shown here follows if we include the correction terms in the summation (2.13) and not merely consider the limit $N \to \infty$ (cf. Problem 2.1). The solid lines in the two panels are calculated using

$$\langle \cos \vartheta \rangle = \frac{\int_0^{2\pi} d\vartheta \, \cos \vartheta \, \exp[-\beta u_{torsion}(\vartheta)]}{\int_0^{2\pi} d\vartheta \, \exp[-\beta u_{torsion}(\vartheta)]}. \tag{2.17}$$

Here $\beta = 1/(k_B T)$, where k_B is Boltzmann's constant and T is the temperature. The torsion potential $u_{torsion}$ is modelled via an empirical potential energy describing butane (see [1]), i.e.

$$\begin{aligned} u_{torsion} \equiv u^B = &355.1(\cos(\vartheta) + 1) \\ &- 68.21(1 - \cos(2\vartheta)) + 791.6(\cos(3\vartheta) + 1). \end{aligned} \tag{2.18}$$

As before $\phi = 112°$. The potential (2.18) is depicted in Fig. 2.7. Note the three pronounced minima. The one in the center (trans) occurs when the C-atoms form a flat zig-zag in the plane. The other two minima are called gauche minus and gauche plus. Note that for this type of hindered rotation C_∞ is significantly larger than before. However, this depends on temperature. Increasing T lets the PE model approach the limit of the freely rotating chain. There is one more line, the long-dashed line, in the bottom panel of Fig. 2.6. Its calculation is based on an approximation which takes into account the coupling between successive torsion angles. We shall discuss this approximation below.

Aside from the contour length L and the mean square end-to-end distance $\langle R_N^2 \rangle$ there are two additional lengths, which are mentioned with equal frequency in the context of polymer chain conformations. The first is the **Kuhn length** κ, which is defined via

$$\kappa \, n = L = b \sin(\phi/2) N \tag{2.19}$$

Fig. 2.6 Top: C_N versus N for various chain models. Bottom: Temperature dependence of C_∞

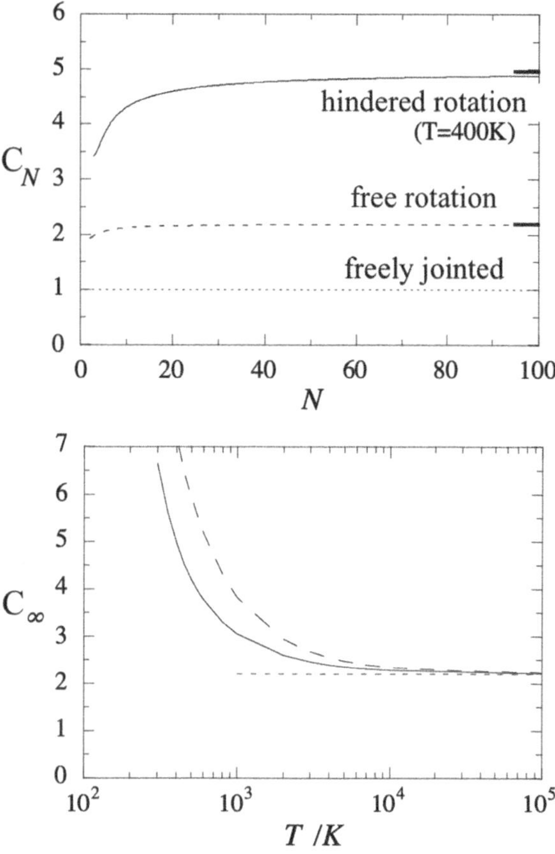

Fig. 2.7 The torsion potential u^B (in units of Kelvin) modelled on the basis of the butane molecule

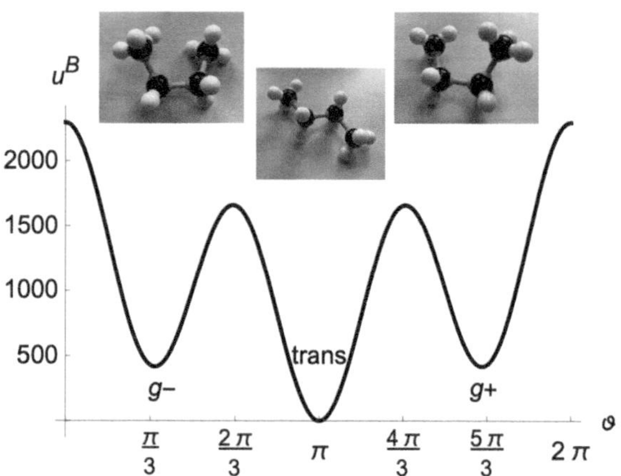

and

$$\kappa^2 n = \langle R^2 \rangle = C_\infty b^2 N. \tag{2.20}$$

This means that a chain is made of n freely jointed (i.e. $\langle \vec{e}_i \cdot \vec{e}_j \rangle = 0$) straight elements, each one Kuhn length long, which possess the same contour length and the same mean-square end-to-end distance as our polymer chain consisting of N monomers of length b. Note that N is considered large and thus C_N is replaced by C_∞. Hence

$$\kappa = \frac{C_\infty}{\sin(\phi/2)} b. \tag{2.21}$$

Since C_∞ usually is between 5 and 10 and $\sin(\phi/2)$ is between 0.8 and 1, κ corresponds to several monomers. One explicit example is our PE-model. At $T = 413\,\mathrm{K}$ we have $C_\infty \approx 7$ (here we use the dashed line in the bottom panel of Fig. 2.6). Since $b \approx 1.5\,\text{Å}$ and $\phi = 112°$, we find $\kappa \approx 1.3$ nm.

The second length is the **persistence length** P. There are polymers, for instance helical ones (like DNA), whose flexibility cannot be described by rotations around bonds. Their conformations are better described by homogeneous bending of the helical backbone as illustrated by the example in Fig. 2.8. The polymer is described as a smooth curve in space, whose local orientation is defined by tangent vectors $\vec{t}$ along this curve. The correlation function $\langle \vec{t}(i) \cdot \vec{t}(j) \rangle$ is given by

$$\langle \vec{t}(i) \cdot \vec{t}(j) \rangle = \exp(-b|i - j|/P), \tag{2.22}$$

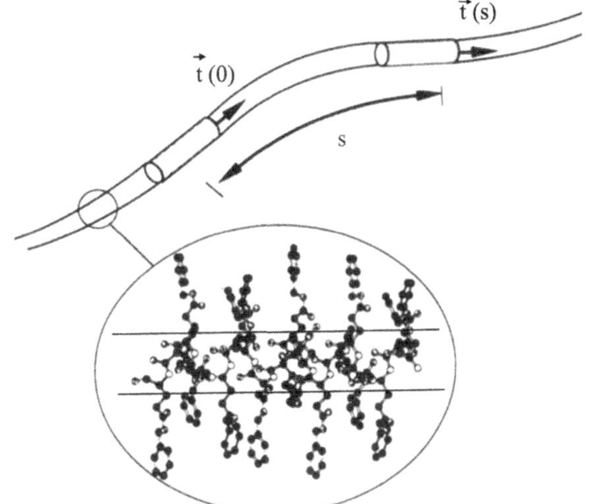

Fig. 2.8 Illustration of a poly(γ-benzyl-L-glutamate) (PBLG) molecule as a homogeneously bend-elastic rod. PBLG is a helical polypeptide enveloped in benzene-terminated side chains for better solubility. Here $\vec{t}(0)$ and $\vec{t}(s)$ are unit tangent vectors, separated by the distance s, along the helical backbone of PBLG

which defines P. Here b is a certain suitable unit of length, e.g., the rise of the helix per turn, and i and j are integers, enumerating the turns of the helix. The contour length of such a polymer is $L = bN$, where N is the number of turns of the helix.

Note that (2.22) is a special case of (2.9). Again $\mathbf{t}$ is given by (2.11). But now $\langle \sin \vartheta \rangle = \langle \cos \vartheta \rangle = 0$. In addition, the angle θ defined via $\vec{t}(i) \cdot \vec{t}(i+1) = \cos \theta$ is related to ϕ via $\theta = \pi - \phi$. Since θ is very small, we use the approximations $\langle \sin \phi \rangle = \langle \sin \theta \rangle \approx 0$ and $-\langle \cos \phi \rangle = \langle \cos \theta \rangle \approx 1 - \frac{1}{2}\langle \theta^2 \rangle$. Hence

$$\langle \mathbf{t} \rangle \approx \begin{pmatrix} 1 - \frac{1}{2}\langle \theta^2 \rangle & 0 & 0 \\ 0 & 0 & 0 \\ 0 & 0 & 0 \end{pmatrix} \tag{2.23}$$

and

$$\langle \vec{t}(i) \cdot \vec{t}(j) \rangle = \left(1 - \frac{1}{2}\langle \theta^2 \rangle \right)^{|i-j|} = \exp(-b|i-j|/P), \tag{2.24}$$

where

$$P = 2b/\langle \theta^2 \rangle. \tag{2.25}$$

We shall return to (2.22) and its derivation in Sect. 5.3.

Let's compute $\langle \vec{R}_N^2 \rangle$ via (2.4) replacing $\langle \vec{e}_i \cdot \vec{e}_j \rangle$ with $\langle \vec{t}(i) \cdot \vec{t}(j) \rangle$. The double sum becomes

$$\sum_{i>j} \langle \vec{t}(i) \cdot \vec{t}(j) \rangle = \sum_{i>j}^{N} q^{i-j} = N\frac{q}{1-q} - q\frac{1-q^N}{(1-q)^2}, \tag{2.26}$$

where $q \equiv \exp(-b/P) < 1$. If we concentrate on the large N limit, i.e. we neglect the second term on the right, then

$$\langle \vec{R}_N^2 \rangle = b^2 \frac{1+q}{1-q} \approx 2LP. \tag{2.27}$$

Note that $b \ll P$ and therefore $q \approx 1 - b/P$. Since in the case of the Kuhn length our result was $\langle \vec{R}^2 \rangle = L\kappa$, we may conclude

$$\kappa = 2P. \tag{2.28}$$

This result is not really a deep interconnection between the Kuhn length and the persistence length. It merely reflects the fact that independent of the flexibility mechanism, i.e. rotation around bonds or continuous bending, an ideal chain becomes a random path if its contour length is much longer than a certain characteristic length—and different characteristic lengths are related.

It is important to bear in mind that κ-values computed via (2.28) exceed those calculated via (2.21)—often by a factor of 100. This is because the two formulas apply to very different flexibility mechanisms.

The above type of chain usually is referred to as **wormlike chain** and sometimes as **persistent flexible polymer**. Contrary to the freely jointed chain, however, which ceases to be useful in the limit $L \sim \kappa$, i.e. when it is so short that its contour length is roughly its Kuhn length, the wormlike chain, which is then called **semi-flexible** rather than wormlike, has numerous applications in the corresponding limit $L \sim P$. This is because persistent flexible polymers in solution may undergo orientation phase transitions, e.g., from an isotropic phase to a nematic phase, dependent on concentration. How this particular **lyotropic liquid crystalline phase transitions** depends on the ratio L/P is discussed in a nice paper by Odijk [2]. In addition, it is worth mentioning that there are important classes of such polymers which are not helical polypeptides. Examples are non-covalent assemblies of globular proteins like **sickle cell hemoglobin** or actin into superstructures. **Tobacco mosaic virus** is another such assembly. These self-assembled polymers are rodlike, i.e. their persistence length is comparable or larger than their contour length. We shall discuss lyotropic liquid crystalline polymers in Sect. 5.3.

2.2   Statistical Thermodynamics of Single Chains

The **conformational partition function** of a polymer chain is given by

$$Q = \sum_p \exp[-\beta U(p)]. \tag{2.29}$$

Here p is an index identifying a particular chain conformation, i.e. the sum extends over all possible conformations of the chain. In practice p is replaced by the cartesian coordinates of the chain's atoms or by the internal coordinates (bonds lengths, valence and torsion angles). U is the potential energy of the chain and $\beta = (k_B T)^{-1}$. For an arbitrary potential energy the evaluation of the sum is very difficult. Here we focus on two summation methods, each of which is based on certain approximations of U. The first method, the **transfer matrix method**, is mainly used to calculate chain properties like its characteristic ratio or its scattering structure factor based on a chemically realistic representation of the monomer units. The second method, the **self-consistent field method**, is useful for the computation of the configurational chain properties if the chain is interacting with an external field (e.g., the density profile of an adsorbed chain) or if the field represents self-interaction of the chain (hence the method's name).

Rotational Isomeric State Model and Transfer Matrix Method:

The transfer matrix method assumes that U includes the internal potential energy of the monomers individually plus the coupling of neighboring monomers. It is also assumed that the relative orientation of a monomer with respect to its neighbors is well described by a small number of discrete (torsion) angles. We discuss the method based on the simple example of the polyethylene (PE) chain with fixed bond lengths and fixed valence angles, i.e. different conformations are entirely due to the variable C-C-torsion angles ϑ.

The conformation potential energy in this case is

$$ U = \sum_{i=1}^{N} u(\vartheta_i, \vartheta_{i+1}) = \sum_{i=1}^{N} \left(u^B(\vartheta_i) + u^P(\vartheta_i, \vartheta_{i+1}) \right). \tag{2.30} $$

The meaning of this expression is explained in Fig. 2.9. Each monomer is defined via a torsion angle ϑ, which in turn is defined by four C atoms or four methylene units respectively (cf. Fig. 2.4). Note that n-butane contains four carbons bonded like this and thus the superscript B. u^B is the same as $u_{torsion}$ in (2.18). The coupling of adjacent monomers arises due to the interaction of the methylene units 1 and 5 in the figure separated by the distance R_4. This coupling is referred to as **pentane effect** and thus the superscript P. Here the **rotational isomeric state (RIS) approximation** is the assertion that every torsion angle assumes one of three discrete values called g_- (gauche minus), t (trans), and g_+ (gauche plus). For the moment this is sufficient.

Figure 2.10 illustrates how we can use the three torsion angle values to express each conformation as a path on a lattice whose rows are ϑ_{g_-}, ϑ_t, and ϑ_{g_+} and whose columns are numbered according to the torsion angles along the chain. By translating each line connecting two dots into a Boltzmann factor we obtain a product of N Boltzmann factors for each conformation, where N is the number of torsion angles along the chain. The subsequent summation over all distinct paths, i.e. products, yields Q.

However, there is a much simpler method. This method is illustrated in Fig. 2.11. For the sake of transparency we assume that every torsion angle can be in one of two states, i.e. 1 and 2 (e.g., ϑ_{g_-} and ϑ_{g_+}). In addition the current chain consists of three monomers only. Now consider two 2×2 **transfer matrices a** and **b**. For instance, the

Fig. 2.9 PE model

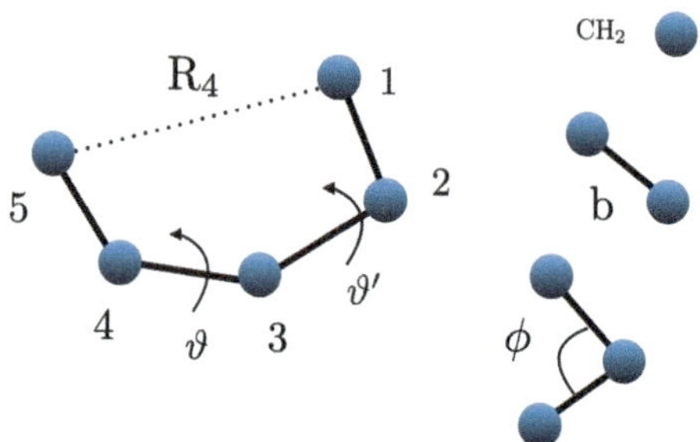

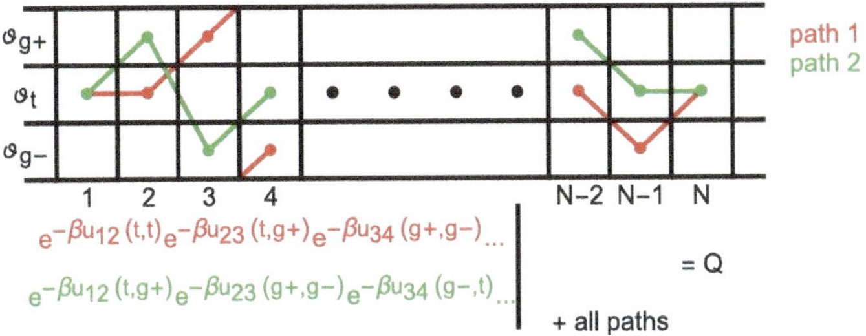

Fig. 2.10 Q as a sum over different paths defined by the torsion angles along the chain. Two paths, one green and one red, are shown. At the bottom we see their (partial) representation as Boltzmann factors in the partition function

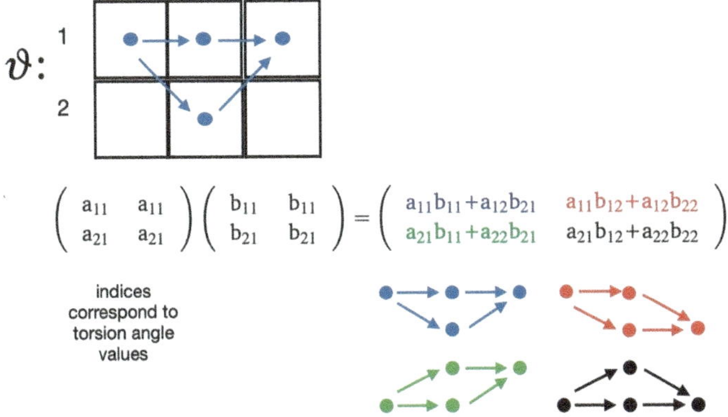

Fig. 2.11 Paths expressed via products of two 2×2 'transfer matrices' **a** and **b**

element a_{11} means that this matrix element takes us from the first monomer/torsion angle in the chain, which is in state 1, to the second monomer/torsion angle in the chain, which is in state 1 also. Analogous, the element b_{12} means that this matrix element takes us from the second monomer/torsion angle in the chain, which is in state 1 again, to the third monomer/torsion angle in the chain, which is in state 2. This particular path is the upper one of the two red paths starting both in state 1 and ending both in state 2. Thus, every element of the product matrix **a b** is the mathematical representation of two paths starting at one particular state and ending in the same or in the other state (cf. the diagrams depicted in the lower right of Fig. 2.11). Thus, we need to figure out what the transfer matrices are in the case of PE, take their N-product, and then add up the elements of the resulting matrix. This does not sound like a real simplification—or does it?

We begin by rewriting Q as follows

$$Q = \sum_{\vartheta_1, \vartheta_2, \ldots, \vartheta_N} \exp\left[-\beta \sum_{i=1}^{N} \left(u^B(\vartheta_i) + u^P(\vartheta_i, \vartheta_{i+1})\right)\right]$$

$$= \sum_{\vartheta_1, \vartheta_2, \ldots, \vartheta_N} \exp\left[-\beta\left(\frac{1}{2}u^B(\vartheta_1) + u^P(\vartheta_1, \vartheta_2) + \frac{1}{2}u^B(\vartheta_2)\right)\right]$$

$$\times \exp\left[-\beta\left(\frac{1}{2}u^B(\vartheta_2) + u^P(\vartheta_2, \vartheta_3) + \frac{1}{2}u^B(\vartheta_3)\right)\right]$$

$$\cdots$$

$$\times \exp\left[-\beta\left(\frac{1}{2}u^B(\vartheta_{N-1}) + u^P(\vartheta_{N-1}, \vartheta_N) + \frac{1}{2}u^B(\vartheta_N)\right)\right]$$

$$\times \exp\left[-\beta\left(\frac{1}{2}u^B(\vartheta_N) + \frac{1}{2}u^B(\vartheta_1)\right)\right].$$

Note that all factors, save for the last one, are of the form

$$T_{ij} = \exp\left[-\beta\left(\frac{1}{2}u^B(\vartheta_i) + u^P(\vartheta_i, \vartheta_j) + \frac{1}{2}u^B(\vartheta_j)\right)\right]. \tag{2.31}$$

The last factor contains no coupling term (If it did, the chain would form a loop as depicted in Fig. 1.4b.) and thus has the form

$$T_{o,ij} = \exp\left[-\beta\left(\frac{1}{2}u^B(\vartheta_i) + \frac{1}{2}u^B(\vartheta_j)\right)\right]. \tag{2.32}$$

The indices i and j label the isomeric states g_-, t, and g_+ in the two neighboring monomers. Hence the corresponding matrix $\mathbf{T}$ has the same meaning as the above matrices $\mathbf{a}$ and $\mathbf{b}$. The meaning of $\mathbf{T_o}$ is analogous—except for one peculiar twist.

The example in Fig. 2.11 taught us to work out the product $\mathbf{T}^{N-1}\mathbf{T_o}$ and then compute the sum over all elements to obtain Q. But the example also included paths which start in one state and end in another. This is not true now. Due to the specific way in which we have rewritten Q, particularly by shifting the factor $\exp\left[-\beta u^B(\vartheta_1)/2\right]$ to the right end of the product of Boltzmann factors, every path terminates in the same state from which it started! If we only include such paths in the example in Fig. 2.11 this mean that instead of summing over all elements of $\mathbf{a}\,\mathbf{b}$ we only sum over the elements along the diagonal of $\mathbf{a}\,\mathbf{b}$, i.e. we calculate the **trace** $\mathrm{Tr}(\mathbf{a}\,\mathbf{b})$—and the trace has very useful properties. Here this means

$$Q = \mathrm{Tr}(\mathbf{T}^{N-1}\mathbf{T_o}), \tag{2.33}$$

which actually is quite easy to calculate.

Let us assume that $\mathbf{T}$ can be diagonalized, i.e.

$$\mathbf{S}^{-1}\mathbf{T}\mathbf{S} = \begin{pmatrix} \lambda_> & 0 & 0 \\ 0 & \lambda_< & 0 \\ 0 & 0 & \lambda_{<<} \end{pmatrix} \equiv \mathbf{D}, \tag{2.34}$$

where $\mathbf{S}\mathbf{S}^{-1} = \mathbf{I}$ and $\mathbf{I}$ is the unit matrix. Hence

$$\begin{aligned} Q &= \mathrm{Tr}(\mathbf{T}^{N-1}\mathbf{T_0}) = \mathrm{Tr}(\mathbf{T}\mathbf{S}\mathbf{S}^{-1}\mathbf{T}\mathbf{S}\mathbf{S}^{-1}\ldots\mathbf{S}\mathbf{S}^{-1}\mathbf{T_0}\mathbf{S}\mathbf{S}^{-1}) \\ &= \mathrm{Tr}(\mathbf{S}^{-1}\mathbf{T}\mathbf{S}\mathbf{S}^{-1}\mathbf{T}\mathbf{S}\mathbf{S}^{-1}\ldots\mathbf{S}\mathbf{S}^{-1}\mathbf{T_0}\mathbf{S}) \\ &= \mathrm{Tr}(\mathbf{D}^{N-1}\tilde{\mathbf{T}}_o) \\ Q &= \lambda_>^{N-1}\tilde{T}_{o,11} + \lambda_<^{N-1}\tilde{T}_{o,22} + \lambda_{<<}^{N-1}\tilde{T}_{o,33}, \end{aligned} \tag{2.35}$$

where

$$\tilde{\mathbf{T}}_o = \mathbf{S}^{-1}\mathbf{T_0}\mathbf{S}. \tag{2.36}$$

Note that the second equation follows via the cyclic permutation of $\mathbf{S}^{-1}$ from left to right, which is one of the useful properties of the trace. With this the (reduced) **conformation free energy** $\beta F = -\ln Q$ of the chain becomes

$$\beta F = -\ln\left(\lambda_>^{N-1}\tilde{T}_{o,11} + \lambda_<^{N-1}\tilde{T}_{o,22} + \lambda_{<<}^{N-1}\tilde{T}_{o,33}\right). \tag{2.37}$$

Let us briefly discuss our notation, i.e. the subscripts $>$, $<$, and $<<$. There is a mathematical theorem, the theorem by Perron–Frobenius, which states the following: Let $\mathbf{A}$ be a positive $N \times N$ matrix, i.e. $A_{ij} > 0\ \forall\ i, j$. This $\mathbf{A}$ possess a real, non-degenerate eigenvalue $\lambda_>$ and $\lambda_> > |\lambda_k|$, where the index k enumerates all other eigenvalues. Here $k = <, <<$, without implying any particular order of these eigenvalues.

Hence we may write the free energy as a sum over three contributions (for sufficiently large N)

$$\beta F = -N\ln\lambda_> - \ln\frac{\tilde{T}_{o,11}}{\lambda_>} + O(e^{-(N-1)/\xi}). \tag{2.38}$$

The first term on the right hand side is the bulk limit dominating for large N. The next term is the surface contribution due to the ends. The third term is a finite size contribution, which vanishes in the limit $N \to \infty$. The quantity ξ is a **correlation length** defined via

$$\xi \equiv -\frac{1}{\ln|\lambda_</\lambda_>|}. \tag{2.39}$$

Essentially, ξ is a measure for how far from the chain ends the latter will have an effect. It also is a measure for the decay of pair-correlations along the chain, e.g., $\langle \vartheta_t(i)\vartheta_t(j)\rangle$, where i and j are two monomers separated by the distance $|i - j|$. However, this is all good and well, but how do we calculate quantities like C_∞?

We start with $p(\vartheta_t; k)$, the probability that the kth torsion angle along the chain is in the trans-state. This quantity is

$$p(\vartheta_t; k) = \frac{Q(\vartheta_t; k)}{Q} = \frac{\sum \text{paths passing through} \vartheta_t \text{ at } k}{\sum \text{all paths}}. \tag{2.40}$$

We do know Q, i.e. the (Boltzmann weighted) sum over all paths. But how do we force a polymer conformation (or path) to be in the trans-state at position k? The answer is that we calculate the partition function as before, except that we insert the matrix

$$\mathbf{I}_{\vartheta_t} = \begin{pmatrix} 0 & 0 & 0 \\ 0 & 1 & 0 \\ 0 & 0 & 0 \end{pmatrix}$$

before the kth transfer matrix in the trace. The result is

$$\begin{pmatrix} 0 & 0 & 0 \\ 0 & 1 & 0 \\ 0 & 0 & 0 \end{pmatrix} \begin{pmatrix} g_- \leftarrow g_- & g_- \leftarrow t & g_- \leftarrow g_+ \\ t \leftarrow g_- & t \leftarrow t & t \leftarrow g_+ \\ g_+ \leftarrow g_- & g_+ \leftarrow t & g_+ \leftarrow g_+ \end{pmatrix} = \begin{pmatrix} 0 & 0 & 0 \\ t \leftarrow g_- & t \leftarrow t & t \leftarrow g_+ \\ 0 & 0 & 0 \end{pmatrix},$$

i.e. we force the path through the trans state at this torsion angle along the chain. Note that here we use the notation $i \leftarrow j$ compared to (2.31) and (2.32). The analogous matrices for the other two probabilities, $p(\vartheta_{g_-}; k)$ and $p(\vartheta_{g_+}; k)$ are

$$\mathbf{I}_{\vartheta_{g_-}} = \begin{pmatrix} 1 & 0 & 0 \\ 0 & 0 & 0 \\ 0 & 0 & 0 \end{pmatrix} \quad \text{and} \quad \mathbf{I}_{\vartheta_{g_+}} = \begin{pmatrix} 0 & 0 & 0 \\ 0 & 0 & 0 \\ 0 & 0 & 1 \end{pmatrix}.$$

Hence

$$p(\vartheta_\alpha; k) = \frac{Tr(\mathbf{T}^{k-1}\mathbf{I}_{\vartheta_\alpha}\mathbf{T}^{N-k}\mathbf{T}_0)}{Tr(\mathbf{T}^{N-1}\mathbf{T}_0)}$$

$$= \frac{Tr(\mathbf{D}^{k-1}\tilde{\mathbf{I}}_{\vartheta_\alpha}\mathbf{D}^{N-k}\tilde{\mathbf{T}}_0)}{Tr(\mathbf{D}^{N-1}\tilde{\mathbf{T}}_0)} \overset{N\to\infty}{=} \tilde{I}_{\vartheta_\alpha,11}. \tag{2.41}$$

Here α stands for either t, g_-, or g_+ and $\tilde{\mathbf{I}}_{\vartheta_\alpha} = \mathbf{S}^{-1}\mathbf{I}_{\vartheta_\alpha}\mathbf{S}$. We also assume that the largest eigenvalue, $\lambda_>$, is the element D_{11}.

Let us now attempt to compute C_∞ for PE with this method. First we need to work out the transfer matrix elements according to (2.31). This means that we need an expression for $u^P(\vartheta, \vartheta')$, since we already have $u_B(\vartheta)$ (cf. (2.18)). $u^P(\vartheta, \vartheta')$ is modelled via

$$u^P(\vartheta, \vartheta') = 4\epsilon_{CC}\left[\left(\frac{\sigma_{CC}}{R_4}\right)^{12} - \left(\frac{\sigma_{CC}}{R_4}\right)^6\right] \tag{2.42}$$

and

$$\begin{aligned}
R_4^2 = 4b^2 + 2b^2\Big\{ &- \cos\phi\left(3 - 2\cos\phi + \cos^2\phi - \sin^2\phi\cos\vartheta\right) \\
&- \sin^2\phi\left(\cos\vartheta + \cos\vartheta' - \sin\vartheta\sin\vartheta'\right) \\
&+ \sin\phi\sin(2\phi)\cos^2\frac{\vartheta}{2}\cos\vartheta'\Big\}.
\end{aligned}$$

Note that $b = 1.53$ Å in the case of the C-C bond length, $\phi = 112°$, $\epsilon_{CC}/k_B = 3.725$ K and $\sigma_{CC} = 4$ Å. This is a steric interaction in the form of a **Lennard-Jones potential** of effective C-atoms, which are separated by four bonds along the chain. The direct separation (after some trigonometry) between these atoms is R_4. Note also that we never mention any of the hydrogen atoms. This is because this is a so-called **united-atom model** and hydrogens are fused with their carbon atom into one effective carbon atom.

Table 2.1 lists the minima of u^B defining our RIS angles. The next table, Table 2.2, shows R_4/σ_{CC}, for the different combination of two successive RIS angles. Values less than unity mean that u^P adds an extra repulsion. Otherwise u^P tends to make this particular combination of torsion angles more favourable. Note that the pairs $\vartheta_{g-}\,\vartheta_{g+}$ and $\vartheta_{g+}\,\vartheta_{g-}$ give rise to a pronounced steric hinderance, which in turn suppresses the likelihood of this pairing. Finally, (2.43) is the transfer matrix based on these values for the RIS angles. At $T = 400$ K it possess the eigenvalues 1.59, 0.32, and -0.22. In this case $\xi \approx 1$, i.e. the effect of the chain ends or the extend of correlations along the chain are short-ranged.

Table 2.1 Minima of u^B defining our RIS angles

		$u^B[K]$
ϑ_{g-}	1.107	417.5
ϑ_t	π	0
ϑ_{g+}	5.176	417.5

Table 2.2 R_4/σ_{CC} for the different combination of two successive RIS angles

	ϑ_{g-}	ϑ_t	ϑ_{g+}
ϑ_{g-}	0.94	1.12	0.70
ϑ_t	1.12	1.27	1.12
ϑ_{g+}	0.70	1.12	0.94

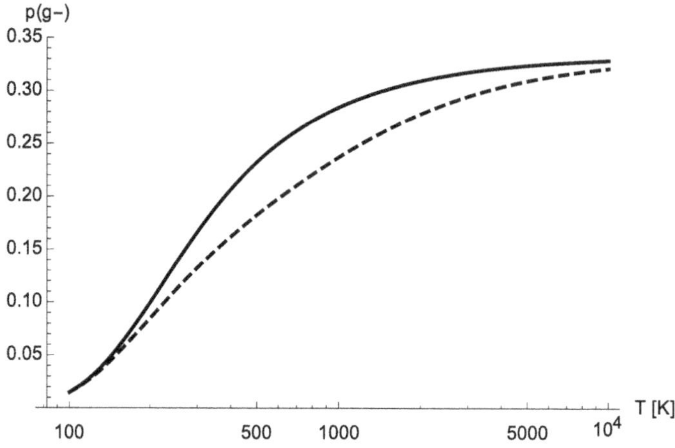

Fig. 2.12 Comparison of $p(\vartheta_{g_-})$ versus T with $u^P = 0$ (solid line) and with $u^P \neq 0$ (dashed line)

$$
\mathbf{T} = \begin{pmatrix} e^{-427/T} & e^{-205/T} & e^{-1468/T} \\ e^{-205/T} & e^{2.72/T} & e^{-205/T} \\ e^{-1468/T} & e^{-205/T} & e^{-427/T} \end{pmatrix} \tag{2.43}
$$

Figure 2.12 shows a comparison of $p(\vartheta_{g_-})$ versus T with $u^P = 0$ (solid line) and with $u^P \neq 0$ (dashed line). Here $p(\vartheta_{g_-})$ is computed according to (2.41) for $N = \infty$. We do not need to also show $p(\vartheta_{g_+})$ and $p(\vartheta_t)$, because $p(\vartheta_{g_-}) = p(\vartheta_{g_+})$ and $p(\vartheta_t) + p(\vartheta_{g_-}) + p(\vartheta_{g_+}) = 1$. We can see that including the u^P-interaction between neighboring monomers makes 'kinks' less likely and therefore we expect that it increases C_∞. We obtain C_∞ using $\langle f(\vartheta) \rangle = \sum_\alpha f(\vartheta_\alpha) p(\vartheta_\alpha)$, where $f(\vartheta)$ stands for $\sin(\vartheta)$ or $\cos(\vartheta)$. Thus in this particular case $\langle \cos \vartheta \rangle = 2(\cos \vartheta_{g_-} + 1) p(\vartheta_{g_-}) - 1$ and $\langle \sin \vartheta \rangle = 0$. Therefore we may use (2.16) and we obtain

$$
C_\infty = \frac{1 - \cos \phi}{1 + \cos \phi} \left(\frac{1}{(\cos \vartheta_{g_-} + 1) p(\vartheta_{g_-})} - 1 \right), \tag{2.44}
$$

which is plotted in Fig. 2.13 (selected values: $T = 100\,\mathrm{K}$, $C_\infty = 101$; $T = 400\,\mathrm{K}$, $C_\infty = 7.2$; $T = 450\,\mathrm{K}$, $C_\infty = 6.6$). Apparently this result is very close to what is shown in the bottom panel of Fig. 2.6. However, a couple of points are worth commenting on. (i) The solid line Fig. 2.6 is computed by averaging $\cos \vartheta$ over the full interval $0 \leq \vartheta \leq 2\pi$, whereas the solid line in Fig. 2.13 is computed by averaging $\cos \vartheta$ over only the three RIS angles. In principle, we may view this as confirmation of the quality of this approximation. (ii) the calculation of the RIS-transfer matrix result in Fig. 2.6 deviates in some (minor) aspects from our above calculation (for details see [3]). Finally we may ask whether our calculation of C_∞ can be verified by experiments? This is indeed possible, using for instance small angle X-ray or neutron scattering. We shall come to this.

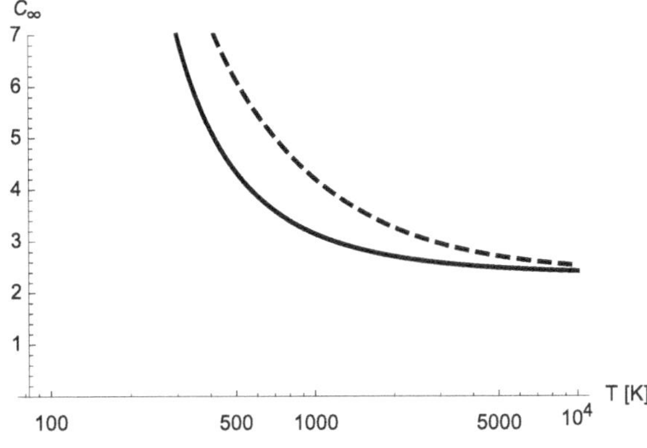

Fig. 2.13 Comparison of C_∞ versus T with $u^P = 0$ (solid line) and with $u^P \neq 0$ (dashed line)

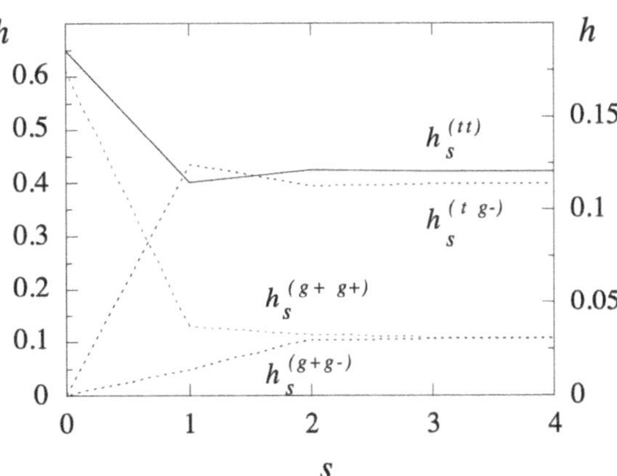

Fig. 2.14 Probability of pairs tt, tg_-, g_-g_+ and g_+g_+ as function of the distance s at $T = 400$ K. The left axis is for the solid lines, while the right axis is for the dashed lines

Before leaving this topic, we briefly want to comment on correlations between the RIS torsion angles. Figure 2.14 shows the probabilities of pairs of torsion angles $h_s^{(\alpha\beta)} \propto \langle \vartheta_\alpha(0)\vartheta_\alpha(s)\rangle$ as function of the separation s. Here α and β stand for g_-, t, and g_+. These calculation were done in the bulk part of the chain, i.e. the chain ends do not affect the results. Technically the calculation is very similar to our above calculation of $p(\vartheta_\alpha)$. We only need to insert two matrices $\mathbf{I}_{\vartheta_\alpha}$ and $\mathbf{I}_{\vartheta_\beta}$ in the proper positions within the product of transfer matrixes—instead just one. The resulting correlations possess a short range. Already after two monomers the result is $h_s^{(\alpha\beta)} = p(\vartheta_\alpha)p(\vartheta_\beta)$. Note that $h_s^{(g_+g_-)} = 0$ for $s = 0$, because one and the same torsion angle cannot be in two different RIS states. The details of this calculation can be found in chapter 8 of [3].

The RIS/transfer matrix method has been applied to almost every polymer. Much of this work was done by Paul Flory and/or his coworkers. Most of the sometimes complicated details are described in his classic text [4]. The first hint I have found that the transfer matrix method is useful for polymers is footnote 5 (referencing

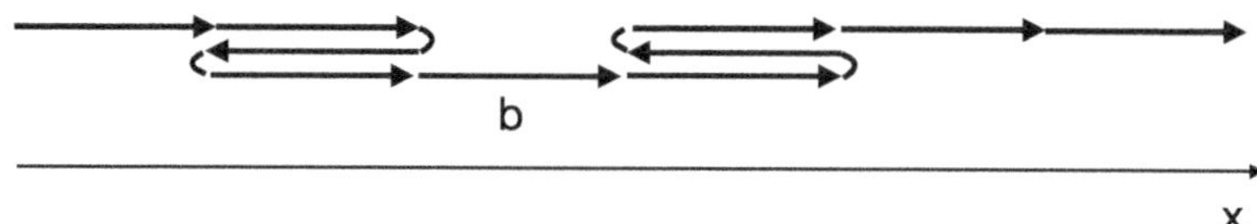

Fig. 2.15 A one-dimensional polymer path along the x-axis

work by Mr. E. Montroll) in a classic paper in the context of the solution of the two-dimensional Ising model with this method by Kramers and Wannier [5].

Self-Consistent Field Method:

This method applies to polymers whose partition function factorises as in the case of the transfer matrix method. For the sake of simplicity we consider a one-dimensional polymer chain depicted in Fig. 2.15. Each Kuhn segment has length b (I sometimes prefer b for the Kuhn length even though b is also used for the monomer length. However, the meaning of b should always be clear from the context.) and can be oriented in either x or $-x$ direction. Therefore there are 2^n possible conformations for a polymer consisting of n Kuhn segments. Let $Q_n(x_1, x_n)$ be the partition function of such polymer starting at x_1 and ending at x_n. The factor 2^n is introduced as a normalization. The Kuhn segments 'feel' the external potential $u(x_i)$ $(i = 1, \ldots, n)$. Hence

$$Q_{n+1}(x_1, x_{n+1}) = \frac{1}{2}\sum_{x_n}{}' Q_n(x_1, x_n)e^{-\beta u(x_{n+1})}. \tag{2.45}$$

$\sum_{x_n}'$ is the sum over all x_n from where we can reach a $(n + 1)$th segment at position x_{n+1}.

At this point we assume the following: (i) $\beta u(x_i) \ll 1\ \forall\ i$; (ii) $u(x)$ varies little on the scale of one Kuhn segment. These assumptions allow us to expand the right hand side of (2.45) at x_{n+1}, i.e.

$$
\begin{aligned}
Q_{n+1}(x_1, x_{n+1}) = \frac{1}{2}\sum_{x_n}{}' \Bigg\{ & Q_n(x_1, x_{n+1}) \\
& + (x_n - x_{n+1}) \left.\frac{\partial Q_n(x_1, x)}{\partial x}\right|_{x_{n+1}} \\
& + \frac{1}{2}(x_n - x_{n+1})^2 \left.\frac{\partial^2 Q_n(x_1, x)}{\partial x^2}\right|_{x_{n+1}} + \cdots \Bigg\} \\
& \times \Big\{ 1 - \beta u(x_{n+1}) + \cdots \Big\}.
\end{aligned}
\tag{2.46}
$$

Using $\sum'_{x_n}(x_n - x_{n+1}) = 0$ and keeping the leading terms only yields

$$Q_{n+1}(x_1, x_{n+1}) - Q_n(x_1, x_{n+1}) \cong \frac{1}{2}b^2 \frac{\partial^2 Q_n(x_1, x)}{\partial x^2}\bigg|_{x_{n+1}}$$
$$-\beta u(x_{n+1})Q_n(x_1, x_{n+1}). \qquad (2.47)$$

Large n allow us to simply use in the continuum limit, i.e.

$$\frac{\partial}{\partial n}Q_n(x', x) = \frac{b^2}{2}\frac{\partial^2}{\partial x^2}Q_n(x', x) - \beta u(x)Q_n(x', x). \qquad (2.48)$$

We solve this partial differential equation employing the ansatz

$$Q_n(x', x) \cong e^{-n\mu_0}\psi_0(x')\psi_0(x). \qquad (2.49)$$

This can be justified in various ways. Note first that (2.48) looks a lot like
Schrödinger's equation in quantum mechanics if n is replaced by a complex time.
From this angle (2.49) looks much like the step that takes us from the time-dependent
Schrödinger equation to the stationary Schrödinger equation. Equation (2.49) can
also be understood as the leading term in an expansion in eigenfunctions. Keeping
the first term only is equivalent to the assumption of **ground state dominance** for
large n. Another angle is this: (i) the free energy and therefore $-k_B T \ln Q_n$ must be
extensive in n; (ii) segments at x' and x are to good approximation independent and
their contributions should therefore factorise. Hence we see that

$$\frac{\Delta F}{nk_B T} \cong -\frac{1}{n}\ln\left[e^{-n\mu_0}\right] = \mu_0 \qquad (2.50)$$

for large n, where $\Delta F = F - F^{(ideal)}$ with $F^{(ideal)}/(nk_B T) = -\ln 2$.
  In any case, inserting (2.49) into (2.48) yields

$$\frac{b^2}{2}\frac{d^2}{dx^2}\psi_0(x) + \left(\mu_0 - \beta u(x)\right)\psi_0(x) = 0, \qquad (2.51)$$

from which we obtain $\psi_0(x)$. But what is the meaning of $\psi_0(x)$? This becomes clear
if we write down the probability for finding a polymer segment at x in terms of the
partition function, i.e.

$$c(x) = \frac{1}{n}\frac{\sum_{x',x'',n'}Q_{n'}(x', x)Q_{n-n'}(x, x'')}{\sum_{x',x''}Q_n(x', x'')}. \qquad (2.52)$$

Here the numerator is the number of paths or polymer conformations starting from (all
possible) x', which, after reaching x in (all possible) n' steps or segments, continue in
$n - n'$ steps to (all possible) x''. The denominator counts all possible paths of length

n connecting (all possible) x' with (all possible) x''. Note that $c(x)$ also is a measure for the polymer concentration at x. Inserting (2.49) into (2.52) yields

$$c(x) \cong \psi_0^2(x). \tag{2.53}$$

In the following we want to demonstrate the use of (2.50), (2.51), and (2.53) via two examples:

• A polymer close to a sticky wall

We assume that a polymer sticks to a wall at position $x = 0$. However, the interaction range is short (on the order of one segment length) and polymer segments outside this distance from the wall are free. We are interested in the polymer concentration outside the range of the direct interaction with the wall, which means that we must solve

$$\frac{b^2}{2}\frac{d^2}{dx^2}\psi_0(x) + \mu_0\psi_0(x) = 0 \tag{2.54}$$

for $\psi_0(x)$. Since we consider a bound state, $\mu_0 < 0$ and the solution far from the wall is given by

$$\psi_0(x) \propto \exp\left[-\kappa x\right], \tag{2.55}$$

where

$$\kappa^2 = -\frac{2\mu_0}{b^2}. \tag{2.56}$$

The attendant change in free energy of the chain due to the adsorption is

$$\frac{\Delta F}{nk_B T} = -\frac{(b\kappa)^2}{2}. \tag{2.57}$$

• A polymer confined inside a one-dimensional slit

The polymer is confined between two hard walls inside a slit of width D. This problem is analogous to a quantum particle in an infinite potential well. As before we assume ground state dominance. For ψ_0 we easily find

$$\psi_0 \propto \sin(\pi x/D) \tag{2.58}$$

and for ΔF

$$\frac{\Delta F}{nk_B T} = \frac{1}{2}\frac{\pi^2 b^2}{D^2}. \tag{2.59}$$

We shall return to the results of these two examples below, where we solve the same or similar problems using different methods.

Remark 1: What does this method have to do with 'self-consistency'? Thus far we have treated the segment potential $u(x)$ as a purely external potential. However, we can assume that segments interact with each other, i.e.

$$u(x) \propto c(x) = \psi_0^2(x). \tag{2.60}$$

Equation (2.51) then becomes a non-linear eigenvalue problem

$$-\frac{b^2}{2}\frac{d^2}{dx^2}\psi_0(x) + \beta\upsilon(T)\psi_0^3(x) = \mu_0\psi_0(x). \tag{2.61}$$

Here $\upsilon(T)$ is a function of temperature.

Remark 2: It is easy to generalize this approach to three dimensions. On a cubic lattice (2.48) becomes

$$\frac{\partial}{\partial n}Q_n(\vec{r}',\vec{r}) = \frac{b^2}{6}\vec{\nabla}_{\vec{r}}^2 Q_n(\vec{r}',\vec{r}) - \beta u(\vec{r})Q_n(\vec{r}',\vec{r}). \tag{2.62}$$

Counting Paths for Different R_N—the Conformation Entropy:

When discussing the mean square end-to-end distance $\langle R^2 \rangle$ we have not stated what the underlying probability distribution of R looks like. Here we obtain this distribution as well as the **conformation entropy** of a polymer chain. We want to base the discussion on the convenient and popular polymer model depicted in Fig. 2.16, which is a chain of Kuhn segments of length b on a cubic lattice. The question whether there is a significant reduction of generality implied by the lattice we shall address at the end of this section.

At this point we ask another question: What is the probability $p(R)$ that the two ends, labeled α and ω, do have the separation R? The answer is $p(R) = \Omega(R)/\sum_R \Omega(R)$. The quantity $\Omega(R)$ is the number of different paths of length n originating from the same lattice point and ending at a distance R from this lattice point. The denominator is the sum over all possible paths of length n originating from the same lattice point without constraining the distance between α and ω. Using Boltzmann's famous entropy formula, $S = k_B \ln \Omega$, we have

$$S(R) - S = k_B \ln p(R), \tag{2.63}$$

Fig. 2.16 A polymer chain on a lattice

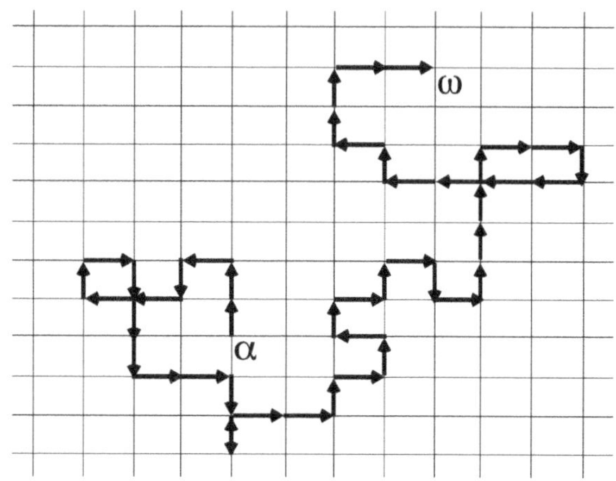

where $S(R) = k_B \ln \Omega(R)$ and $S = k_B \ln \sum_R \Omega(R)$. The end-to-end vector $\vec{R}$ is given by

$$\vec{R} = \sum_{i=1}^{n}(x_i, y_i, z_i) = \left(\sum_{i=1}^{n} x_i, \sum_{i=1}^{n} y_i, \sum_{i=1}^{n} z_i \right), \tag{2.64}$$

where x_i, y_i, and z_i are random variables. Each of these may assume the values $\{-b, 0, 0, 0, 0, b\}$ with equal likelihood. Here b is the lattice spacing and the six values correspond to the six possible orientations of the links (in Fig. 2.16) along the main axes of the cubic lattice.

We obtain $p(R)$ via an important mathematical theorem—the **central limit theorem**. This theorem states that if the s_i are random variables, whose average is μ_s and whose mean square fluctuation is σ_s^2, then the new random variable

$$S_n = \frac{\sum_{i=1}^{n} s_i - n\mu_s}{\sigma_s \sqrt{n}} \tag{2.65}$$

possesses the probability density

$$f(S_n) = \frac{1}{\sqrt{2\pi}} \exp[-S_n^2/2] \tag{2.66}$$

in the limit of infinite n. However, this is a very good approximation even if n is not very large. The reader may confirm this by working through Problem 2.5. Based on the central limit theorem we immediately conclude

$$p(R) \approx \left(\frac{3}{2\pi nb^2}\right)^{1/2} \exp\left[-\frac{3}{2}\frac{R_x^2}{nb^2}\right]$$

$$\times \left(\frac{3}{2\pi nb^2}\right)^{1/2} \exp\left[-\frac{3}{2}\frac{R_y^2}{nb^2}\right]$$

$$\times \left(\frac{3}{2\pi nb^2}\right)^{1/2} \exp\left[-\frac{3}{2}\frac{R_z^2}{nb^2}\right],$$

i.e.

$$p(R) = \left(\frac{3}{2\pi nb^2}\right)^{3/2} \exp\left[-\frac{3}{2}\frac{R^2}{nb^2}\right], \tag{2.67}$$

where $\mu_x = \mu_y = \mu_z = 0$, $\sigma_x^2 = \sigma_y^2 = \sigma_z^2 = b^2/3$, and $4\pi \int_0^\infty dR\,R^2 p(R) = 1$. Using this expression in (2.63) we obtain

$$S(R) = S(0) - \frac{3k_B R^2}{2nb^2}. \tag{2.68}$$

Note that the conformation entropy decreases when the separation of the ends of the chain is increased. This is because stretching a chain reduces the number of paths along which it can join the two endpoints.

Equation (2.67) is the limiting form of $p(R)$ when the polymer chain is purely a random walk and molecular details are not important. Note also that the above results are independent of the constraint of the Kuhn segments to a cubic lattice. Finally, it is worth pointing out that (2.67) as well as (2.68) are frequent building blocks in the various construction schemes for the free energy of polymer systems in the subsequent sections.

2.3 Flory's Calculation of the Exponent ν

We had introduced the exponent ν in (2.8)—the general expression for the mean square end-to-end distance in terms of the chain length. For an ideal chain $\nu = 1/2$. But what is the value of ν when the chain is not ideal—and what does 'not ideal' mean? 'Ideal' meant that chain conformations, on a certain length scale, are random walks. An ideal chain does not interact with itself (beyond short-ranged interactions of the monomers in close proximity)—or other chains for that matter. An intuitive and rather simple calculation, which allows us to study the influence of 'interaction' on ν and thus on the mean square end-to-end distance is this:

Consider a polymer chain occupying a certain volume R^d, where d is the dimension of space. Here R is the same quantity as in (2.68). The free energy of the chain in units of $k_B T$ may be expressed as the sum of two contributions, i.e.

$$f(R) = f^{el}(R) + f^i(R). \tag{2.69}$$

The contribution $f^{el}(R)$ is due to our chain entropy in (2.68), i.e.

$$f^{el}(R) = \frac{3R^2}{2C_\infty b^2 N} + \text{const.} \tag{2.70}$$

Here we substitute $C_\infty b^2 N$ for $b^2 n$.

The interaction term is modelled in analogy to the free energy (per $k_B T$) of a dilute real gas of small molecules, i.e. $f \approx f_{ideal} + N\rho B_2(T)$. The quantity f_{ideal} is the free energy of an ideal gas of N molecules, whose number density is ρ. The quantity $B_2(T)$ is the **second virial coefficient**. Hence

$$f^i(R) = N\rho\upsilon(T), \tag{2.71}$$

where $\rho = N/R^d$ and $\upsilon(T)$, like $B_2(T)$, is a function of T. In other words, the N monomers of the chain are treated as entities analogous to the molecules in a dilute gas.

Combination of the (2.69), (2.70), and (2.71) yields

$$f(R) = \frac{3R^2}{2C_\infty b^2 N} + \upsilon(T)\frac{N^2}{R^d} + \text{const.} \tag{2.72}$$

According to the second law of thermodynamics the free energy wants to be at its minimum. This means that the polymer size R is adjusted to an equilibrium value R_o following from $df(R)/dR|_{R_o} = 0$, i.e.

$$\frac{3R_o}{C_\infty b^2 N} - d\upsilon(T)\frac{N^2}{R_o^{d+1}} = 0$$

or

$$R_o^{d+2} \propto N^3.$$

Since $R_o \propto N^\nu$ we conclude that

$$\nu = \frac{3}{d+2}. \tag{2.73}$$

The ν-values for different d are compiled in Table 2.3. The value for $d = 1$ is without doubt correct. The values for $d = 2$ and $d = 3$ can be compare to the values discussed

Table 2.3 Dimension dependence of ν

d	ν
1	1
2	$3/4 = 0.75$
3	$3/5 = 0.6$
≥ 4	$1/2 = 0.5$

in [6]. When $d = 2$, Flory's value agrees with the exact result for self-avoiding walks on honeycomb lattices. Detailed work also shows that the lattice type should have no influence on the value of ν. The authors also quote experiments yielding $\nu \approx 0.79$. Table 31 in the review contains a list of values favouring $\nu \approx 0.59$ when $d = 3$. These numbers were obtained mainly via renormalization group and Monte Carlo calculations. For $d = 4$, Flory's method again yields the ideal value, $\nu = 0.5$. But what about $d > 4$?[1]

We expect $R_o \geq R_{ideal}$ and thus

$$f^i \leq \upsilon\,(T)\,\frac{N^2}{R_{ideal}^d} \propto \upsilon\,(T)\,N^{2-d/2} \overset{N \to \infty}{\longrightarrow} 0$$

for $d > 4$! Since f^{el} does not vanish equally quick, we conclude that the chain is ideal for $d > 4$.

Our above interaction term is based on the notion of a dilute gas. But when does a polymer chain resemble a dilute gas? The situation in which the monomers inside a polymer chain can be compared to a dilute gas arises when the polymer is in a solvent. This means that the above interaction term $\upsilon(T)$ effectively includes the solvent, but it is still modelled according to the second virial coefficient of a gas. The simplest theory, which nevertheless contains all important ingredients, is the van der Waals theory in which $B_2(T) = b^3 - a/(k_B T)$. Here b^3 and a describe the (excluded) volume of the gas molecules and their long-range attraction, respectively (**van der Waals equation**: $P = \rho k_B T/(1 - \rho b^3) - a\rho^2$, where $\rho = N/V$). It is customary to distinguish different types of solvent in terms of the effective quantities a and b^3:

- Athermal solvents: In an athermal solvent there is only hard-body repulsion. Here this is the limit of the above $B_2(T)$ when T becomes infinite (or $a = 0$) and the molecules are hard bodies. Even though this is never quite realized in nature, the concept is useful because of its simplicity.

- Good solvent: In a good solvent there is a small attraction, reducing the repulsion, and thus the polymer appears 'fluffy'.

---

[1] $d \geq 4$ may seem academic. But in practice, $d = 4$ is often a useful limit.

• θ-solvent: Note that both b^3 and a are positive and thus we can adjust T so that $B_2(T)$, or $\upsilon(T)$ for that matter, is zero. This temperature is the **Boyle temperature** in a gas.

What does this mean for our above calculation if $\upsilon(T_\theta) = 0$? It means we must look at the next term in the expansion in powers of the density when we model f^i. Hence

$$f^i(R) \sim R^d \rho^3 \upsilon'(T) \propto \frac{N^3}{R^{2d}}. \tag{2.74}$$

This leads to

$$\nu = \frac{2}{d+1}. \tag{2.75}$$

In $d = 3$ dimensions we now have $\nu = 1/2$, i.e. the chains are ideal under θ-conditions.

• Poor solvent: The second virial coefficient becomes negative when T is below the Boyle temperature. Analogously, we expect that for $T < T_\theta$ the attractions overwhelm the repulsion, which yields compact conformations. Of course, we must proceed with caution here, because we are looking at just one term in an expansion. But nevertheless, it is plausible that the polymer chain will collapse at some point and essentially precipitate in solution. In this case the monomers are densely packed like molecules inside a liquid droplet.

An experiment in which the temperature is varied across the θ-temperature is shown in Fig. 2.17. The figure depicts the temperature dependence of the radius of gyration (S) and the hydrodynamic radius (R_H), both quantities are measures of the size of a polymer chain akin to the mean square end-to-end distance, for polystyrene $(M_w = 2.6 \cdot 10^7)$ in cyclohexane. The phenomenon is called **coil-globule transition**.

The relation between R_o and N for these types of solvent in three dimensions is summarized in Table 2.4.

2.4 The Scaling Concept

Most of you are familiar with simple examples of **dimensional analysis**. One such example is the time period T_p of a pendulum. In the first step of the dimensional analysis, we compile a list of all physical quantities that we believe have something to do with T_p. These quantities include m, the mass of the weight at the end of the otherwise massless string, l, the length of the string, and g, the gravitational acceleration. In a second step we write down these quantities including their physical dimension (cf. Table 2.5).

Fig. 2.17 Temperature dependence of the radius of gyration (S) and the hydrodynamic radius (R_H) for polystyrene ($M_w = 2.6 \cdot 10^7$) in cyclohexane. Reprinted with permission from [7]

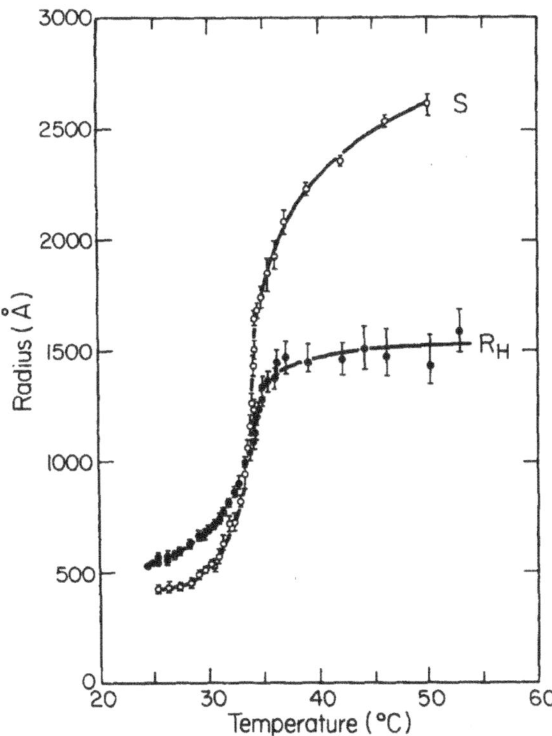

Table 2.4 Size of polymer chain in different solvents

	$R_o \sim$
Athermal solvent	$N^{3/5}$
Good solvent	$N^{3/5}$
θ-solvent	$N^{1/2}$
Poor solvent	$N^{1/3}$

Table 2.5 Quantities potentially affecting the period of a mathematical pendulum including their physical dimensions

T_p	s
m	kg
l	m
g	m s^{-2}

Next we construct dimensionless expressions based the quantities in Table 2.5. Here there is only one:

$$\frac{T_p^2 g}{l} \sim 1. \tag{2.76}$$

The notation ~ 1 means that $T_p^2 g/l$ is (most likely) on the order of one. Note that m does not appear, because the unit kg is not cancelled by any of the other units. Hence we find

$$T_p \sim \sqrt{l/g}, \tag{2.77}$$

which is correct except for a factor 2π on the right hand side.

Essentially, **scaling** is based on the same idea, i.e. identification of the 'basic' quantities in a physical problem and using them to construct dimensionless expressions. The difference is that basic quantities are not merely quantities like b or N. Instead they are slightly more complex. So what are they?

Let's assume we are interested in a polymer's free energy F, which is the equivalent of T_p in the above example. Here F of course is a piece of the polymer's total free energy only. This piece depends on the specific problem. What we always exclude are contributions to the total free energy from the many atomic degrees of freedom (mainly vibrational modes). Now let's talk about the basic quantities F may depend upon.

Figure 2.18 shows a polyethylene chain constructed via an algorithm based on the transfer matrix approach. We shall explain the algorithm in the next section. Here we merely remark that this chain consists of 10^4 monomers. The key observation is that it appears divided into **blobs**—highlighted by the red circles. (A note of caution: Not everything that looks like a blob may in fact be one. Some blobs disappear when

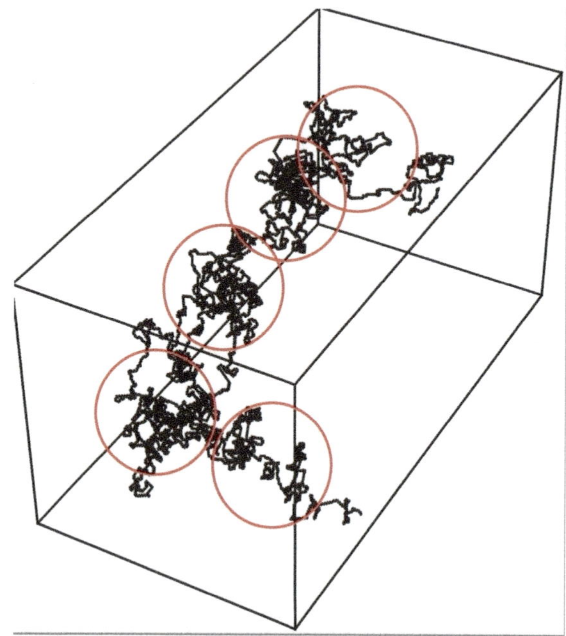

Fig. 2.18 Polyethylene chain consisting of 'blobs'

we look at the chain from a different direction.). Let's assume these blobs all have roughly the same diameter ξ and within each blob there are roughly N_ξ monomers. The two quantities should be related via

$$\xi \sim b N_\xi^\nu. \tag{2.78}$$

This means that ξ is likely one of the aforementioned basic quantities.

Another basic quantity is the typical 'thermal blob energy', which we assume is $k_B T$. The idea is that a rotation with respect to a single bond, corresponding to energies on the order of $k_B T$, is sufficient to significantly alter the shape of a blob. Note again that this does not include the 'many $k_B T$s' from the atomic degrees of freedom, since those presumably do not affect the piece of F we are interested in.

The relation equivalent to (2.76) is

$$\frac{F}{k_B T} \sim \frac{N}{N_\xi}. \tag{2.79}$$

Here we assert that F is extensive in the number of blobs, which is N/N_ξ. In the following three examples we want to practice the application of (2.79).

• Conformation free energy of polymers in dilute solution

Relation (2.79) by itself does not tell us much, because we do not know N_ξ or if we replace N_ξ with ξ/b via (2.78) we do not know ξ. In essence we need one extra equation or relation, allowing to eliminate N_ξ in (2.79). This relation is

$$R \sim \xi \frac{N}{N_\xi}. \tag{2.80}$$

The quantity R of course is the end-to-end distance of the polymer. Note that (2.80) is reasonably well supported by Fig. 2.18.

Using (2.80) in conjunction with (2.78) and (2.79) yields the configuration free energy of the chain:

$$F \sim k_B T \left(\frac{R}{b N^\nu} \right)^{1/(1-\nu)}. \tag{2.81}$$

Setting $\nu = 1/2$, i.e. the polymer chain is ideal, we find that the result is in complete accord with (2.68) (note: $S = -\partial F/\partial T$). This is important, since it confirms that our intuition thus far is correct. It also strengthens our confidence in a new result, which we obtain with $\nu = 3/5$, i.e. the real chain-exponent!

Remark: If we pull on the ends of the chain, we can calculate the magnitude of the restoring force $f^{(el)}$ via $f^{(el)} = dF/dR$, which yields

$$f^{(el)} \sim k_B T \left(\frac{R}{b^{1/\nu} N} \right)^{\nu/(1-\nu)} .$$ (2.82)

This force will be linear (**Hook's law**) for an ideal chain. But in the case of a real chain it will increase with a larger power, i.e. $(R/N)^{3/2}$ for Flory's ν-value in three dimensions. Eventually however, (2.82) must break down when R approaches the contour length. The finite extensibility of the chain, which is not included here, leads to a rapid increase of the force.

• Weakly adsorbed chain

In this example we estimate the free energy of a weakly adsorbed chain. As before we utilize the blob-picture as illustrated in Fig. 2.19. The size of the absorbed blobs is once again given by (2.78) and the free energy of adsorption F_{ad} of the weakly adsorbed chain is once again given by (2.79). The only modification is that $k_B T$ is replaced by $-k_B T$, since this is the (negative) blob adsorption energy. And, as in the previous example, we need one extra relation to eliminate N_ξ. This relation involves δ, the fraction of adsorption energy (in units of $k_B T$) of a single polymer segment inside a layer of thickness b above the surface (i.e. we assume that the range of the interaction with the surface is of this particular size).

We obtain the number of monomers in contact with the surface, i.e. the monomers inside a layer of thickness b, via multiplication of the number density of monomers inside the blob, N_ξ/ξ^3, with the volume of the layer $\xi^2 b$:

$$\frac{N_\xi}{\xi^3} \xi^2 b \sim \left(\frac{b}{\xi} \right)^{1-1/\nu} .$$ (2.83)

Every monomer in this layer has a small fraction δ of the blob's total adsorption energy. Hence

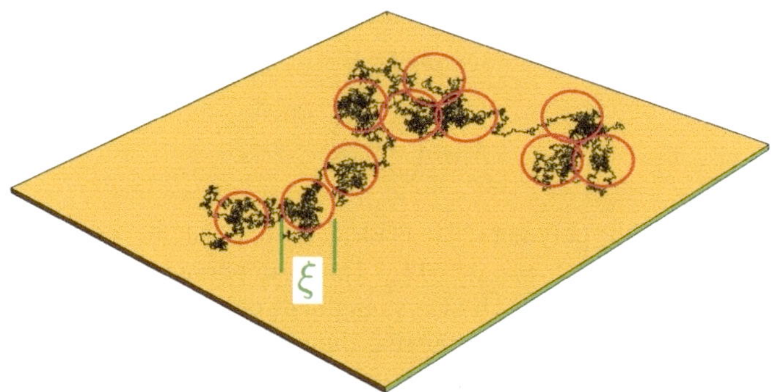

Fig. 2.19 Weakly adsorbed chain in the 'blob'-picture

$$\left(\frac{b}{\xi}\right)^{1-1/\nu} \delta \sim 1, \tag{2.84}$$

which yields

$$\xi \sim b\delta^{\nu/(\nu-1)}, \tag{2.85}$$

the desired additional relation.

We conclude that the free energy of adsorption of the chain is

$$F \sim -k_B T \frac{N}{N_\xi} \approx -k_B T N \delta^{1/(1-\nu)}$$

$$= -k_B T N \begin{cases} \delta^2 & (\nu = 1/2) \\ \delta^{5/2} & (\nu = 3/5) \end{cases}. \tag{2.86}$$

Note that the ideal chain result in this example agrees with the result (2.57) obtained using the self-consistent field approach.

- A polymer confined inside a tube

Finally, we study a polymer confined inside a tube of diameter D. The situation is depicted in Fig. 2.20. This is very similar to the polymer in a one-dimensional slit discussed previously using the self-consistent field approach.

We are interested in the D-dependence of the polymer's configuration free energy. As in the previous two examples we rely on the blob-picture and on (2.78) and (2.79). Here our extra equation is

$$\xi \approx D, \tag{2.87}$$

i.e. the blob-size is constraint by the tube's diameter. Hence

$$F(D) \sim k_B T \frac{N}{N_\xi} \approx k_B T N \begin{cases} \left(\frac{b}{D}\right)^2 & (\nu = 1/2) \\ \left(\frac{b}{D}\right)^{5/3} & (\nu = 3/5) \end{cases}. \tag{2.88}$$

Fig. 2.20 Polymer confined to a tube

Note that F for $v = 1/2$ once again agrees with our previous result in (2.59), obtained using the self-consistent field approach, within a factor (see also Problem 2.6). This difference is not surprising, since (i) the scaling approach (usually) does not yield numerical factors and (ii) (2.59) was obtained in one dimension.

In summary, the scaling approach involves a certain amount of sound physical observation and intuition. Here the observation is that longer chains appear to form chains of blobs. Therefore the blob's linear dimension ξ becomes the basic length. How the chain of blobs behaves under given circumstances is where the intuition comes in. Using $k_B T$ as the basic energy unit is another important ingredient. Yet another is that the free energy is extensive.

Scaling allows to obtain significant results quickly. However, there is the risk that an observation is not sound or the intuition is faulty, which produces incorrect results. Therefore it is always good if a comparison with results obtained by other methods, like the ideal chain results in the above examples, is possible. This then inspires confidence in the new results for real chains.

2.5 Measuring Size and More by Scattering

We have invested considerable effort into the computation of C_∞ (for PE). Therefore we should try and verify our results. This can be done via small angle scattering experiments—for instance using X-rays or neutrons. Visible light also is an option. Note that the 'object's' size can be quite large here, i.e. $\sqrt{\langle R_N^2 \rangle} = \sqrt{C_\infty b^2 N^{2v}} \sim 10$ nm (assuming $C_\infty \sim 10$, $b \sim 5$ Å, $N \sim 10^5$, and $v = 3/5$).

Before we begin our discussion of scattering, we take a moment to introduce another measure of size. Even though we had claimed that the square root of the mean-square end-to-end distance $\langle R_N^2 \rangle$ is a good measure for 'size' or 'linear dimension', this is not always true. For instance, how would we apply $\langle R_N^2 \rangle$ to a branched polymer with many ends? A more general measure for 'size' or 'linear dimension' can be defined via the mean-square **radius of gyration**:

$$\langle R_g^2 \rangle = \frac{1}{N} \sum_{i=1}^{N} \langle (\vec{r}_i - \vec{r}_{cm})^2 \rangle \overset{*}{=} \frac{1}{N^2} \sum_{i>j}^{N} \langle r_{ij}^2 \rangle. \tag{2.89}$$

Here $\vec{r}_{cm} = N^{-1} \sum_{j=1}^{N} \vec{r}_j$ is the center of mass of the polymer. We leave * as an exercise to the reader. If we use once again $\langle r_{ij}^2 \rangle = b^2 |i - j|$, we can easily obtain

$$\langle R_g^2 \rangle = \frac{1}{6} \langle R_N^2 \rangle \tag{2.90}$$

for ideal linear chains (note: $\sum_{i>j}^{N} |i - j| = (N - 1)1 + (N - 2)2 + \cdots = \frac{1}{6}(N^2 - 1)N \approx \frac{1}{6}N^3$).

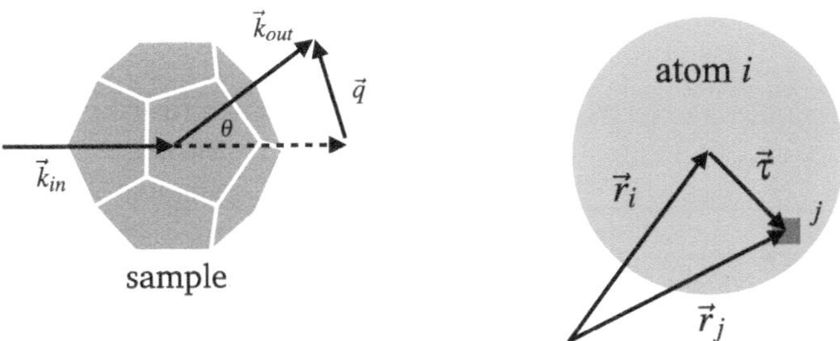

Fig. 2.21 Left: The scattering vector $\vec{q}$ resulting from the wave vectors of incoming and outgoing radiation/particles. Right: Definition of the vectors $\vec{r}_i$ and $\vec{r}_j$ and $\vec{\tau}$ in relation to the scattering center (an atom in this case)

Now let us turn to scattering. The sketch on the left in Fig. 2.21 illustrates the general setup of such experiments. Incoming radiation with momentum $\vec{k}_{in}$ is scattered (elastically) into $\vec{k}_{out}$ by the sample. The momentum transfer or magnitude of the scattering vector $\vec{q} = \vec{k}_{out} - \vec{k}_{in}$ is

$$q = \frac{4\pi}{\lambda}\sin(\theta/2). \tag{2.91}$$

λ is the wavelength of the radiation (note: $k_{in} = k_{out} = 2\pi/\lambda$).

When we discussed light scattering in Chap. 1 we arrived at (1.14) and in particular at

$$P_i(\theta) = \frac{1}{V^2}\int_V d^3r\,d^3r'\,e^{i\vec{q}\cdot(\vec{r}-\vec{r}')}\frac{\langle c_w(\vec{r})c_w(\vec{r}')\rangle}{\langle c_w\rangle^2}. \tag{2.92}$$

While we obtained $P_i(\theta)$ by considering the scattering of visible light, this expression also can be used to obtain the angular distribution of scattered intensity for other types of radiation like **X-rays** or **neutrons**. There are still other probes like low-energy electron diffraction (LEED), which is used to study solid surfaces under vacuum conditions. In this case (2.92) is not sufficient, because we must include multiple scattering, i.e. we must go beyond the 1st Born approximation. But fortunately this is not relevant for the study of polymers with electromagnetic radiation or neutrons.

Let's assume we want to calculate the scattering amplitude collected over a volume element ΔV. In the example on the right hand side in Fig. 2.21 the volume element i at the position $\vec{r}_i$ is occupied by an atom, but the same concept can be applied to a molecule or a monomer or even a group of neighboring monomers within a polymer. In any case, the scattering amplitude is given by

$$\int_{\Delta V} d^3r\,e^{i\vec{q}\cdot\vec{r}}c(\vec{r}) = \int_{\Delta V} d^3\tau\,e^{i\vec{q}\cdot(\vec{r}_i+\vec{\tau})}c(\vec{\tau}) = e^{i\vec{q}\cdot\vec{r}_i}mf(\vec{q}), \tag{2.93}$$

where

$$f_i(\vec{q}) = \frac{1}{m} \int_{\Delta V} d^3\tau \, e^{i\vec{\tau}\cdot\vec{q}} c(\vec{\tau}) \tag{2.94}$$

is a so called **form factor**. The mass m is given by $m = \int_{\Delta V} d^3\tau \, c(\vec{\tau})$. The form factor sums over the scattering amplitudes from within ΔV. In general $c(\vec{\tau})$ is not merely a mass density as in our case -we developed (2.92) based on our discussion of light scattering—but it will be a function of the interaction between the probing radiation/particle beam and the atom, monomer etc. with which the radiation/particle beam interacts. Nevertheless, this interaction is proportional to the density of atoms, monomers etc. within ΔV, and thus our formalism applies to, for instance, neutron scattering. Using the form factor we can rewrite $P_i(\theta)$ as

$$P_i(\theta) = \frac{1}{\langle N \rangle^2} \langle \sum_{i,j=1}^{N} f(\vec{q})f(-\vec{q})e^{i\vec{q}\cdot\vec{r}_{ij}} \rangle, \tag{2.95}$$

where $\vec{r}_{ij} = \vec{r}_i - \vec{r}_j$. N is the number of filled volume elements (let's say monomers) in the system. It is a convenient approximation to set the form factor equal to one, i.e. $f(\vec{q}) \to f(0) = 1$. This is reasonable, because we shall be interested in the scattering from at least an entire polymer coil, which has a much larger linear dimension than the entities producing the form factor. Due to the reciprocal relation between linear dimension in real space versus q-space, $f(\vec{q})$ varies little in the q-range we are interested in. Thus our final formula for $P_i(\theta)$ becomes

$$P_i(\theta) = \frac{1}{\langle N \rangle^2} \langle \sum_{i,j=1}^{N} e^{i\vec{q}\cdot\vec{r}_{ij}} \rangle. \tag{2.96}$$

However, there are situations when instead of summing over the positions of monomers we prefer an integration in the continuum. For this purpose we can represent the monomer number density via

$$\rho(\vec{r}) = \sum_{i=1}^{N} \delta(\vec{r} - \vec{r}_i). \tag{2.97}$$

Note that $\int_V d^3r \dots$ yields the same results, i.e. N, on the two sides of this equation. Hence we obtain the following continuum form of (2.96):

$$P_i(\theta) = \frac{1}{\langle N \rangle^2} \int_V d^3r \, d^3r' \, e^{i\vec{q}\cdot(\vec{r}-\vec{r}')} \langle \rho(\vec{r})\rho(\vec{r}') \rangle. \tag{2.98}$$

In what follows we shall first employ the discrete summation, i.e. (2.96), to obtain the angular dependence of the scattering intensity collected from polymer coils, before returning to the continuum formula.

Let's suppose that we momentarily study a closed system in which the number of monomers is constant, i.e. $N = \langle N \rangle$. The meaning of $\langle \sum_{i,j=1}^{N} \ldots \rangle$ in this case is twofold. Suppose someone rotates our sample. Will this change $P_i(\theta)$? No, because we look at many polymers at the same time and our overall system is isotropic. Thus, one piece of $\langle \sum_{i,j=1}^{N} \ldots \rangle$ is an orientation average, keeping the relative positions (and therefore $r_{ij} = |\vec{r}_{ij}|$) of the scattering centers fixed, i.e.

$$\langle e^{i\vec{q}\cdot\vec{r}_{ij}} \rangle_{orient} = \frac{1}{4\pi} \int_0^{2\pi} d\varphi \int_0^{\pi} d\vartheta \, \sin \vartheta \, e^{iqr_{ij}\cos\vartheta} = \frac{\sin(qr_{ij})}{qr_{ij}}. \tag{2.99}$$

Here we have used the direction of $\vec{q}$ as our z-direction.

Now suppose we scatter from a single chain for a long time or our scattering intensity is independently contributed from many identical chains. Is r_{ij}, where i and j denote monomers either in the single chain or the ith and the jth monomers in each of the many chains, a constant? No, it changes due to conformation changes and our scattering intensity is an average over all possible values of r_{ij} for all pairs ij. Hence,

$$\langle e^{i\vec{q}\cdot\vec{r}_{ij}} \rangle = \langle \frac{\sin(qr_{ij})}{qr_{ij}} \rangle_{conf} = \int d^3r_{ij}\, p(r_{ij}) \frac{\sin(qr_{ij})}{qr_{ij}}, \tag{2.100}$$

where $p(r_{ij})$ is the normalized probability density that the distance between i and j is r_{ij}. But how do we find $p(r_{ij})$?

Scattering from Ideal Chains:

The answer depends on the level of approximation we chose. Equation (2.67) gives $p(r_{ij})$ when i and j are the monomers at the chain's ends. But if i and j are inside the chain we can still use (2.67). Instead of the end-to-end distance R we now substitute the i-to-j distance r_{ij} and n becomes $|i - j|$, i.e.

$$p(r_{ij}) = \left(\frac{3}{2\pi |i - j| b^2} \right)^{3/2} \exp\left[-\frac{3}{2} \frac{r_{ij}^2}{|i - j| b^2} \right]. \tag{2.101}$$

Inserting (2.101) into (2.100) yields

$$\langle e^{i\vec{q}\cdot\vec{r}_{ij}} \rangle = \exp[-|i - j| b^2 q^2 / 6]. \tag{2.102}$$

Hence,

$$P_i(\theta) = \frac{1}{N^2} \sum_{i,j=1}^{N} e^{-|i-j|b^2 q^2/6}$$

$$\overset{*}{=} \frac{1}{N^2} \frac{\left(N(e^{2a} - 1) + 2e^a(e^{-aN} - 1)\right)}{(e^a - 1)^2},$$
(2.103)

where $a = b^2 q^2/6$ (*: here I use Mathematica to do the summation). Note that the summation is analogous to the one in (2.26). When $a \ll 1$ we can use $e^a \approx 1 + a$ and, keeping the leading order only, we obtain

$$P_i(\theta) \approx \frac{2(e^{-x} + x - 1)}{x^2} \rightarrow \begin{cases} \frac{2}{x} & (x \gg 1) \\ 1 - \frac{1}{3}x & (x \ll 1) \end{cases}.$$
(2.104)

Here $x = aN = q^2 \langle R_g^2 \rangle$. This formula was first derived by Debye in 1947. Note that $x \gg 1$ is sensible, despite our previous assumption $a \ll 1$, when N is sufficiently large. The limit $x \ll 1$ is known as **Guinier's law. Debye's formula** and its two limits are depicted in Fig. 2.22. Note that at $x \sim 1$ the intensity exhibits its strongest dependence on x as we had stated above. We may say that Debye's formula, or Guinier's law for that matter, is the form factor of a polymer based on (2.67). In the limit $x \ll 1$ or $q^2 \langle R_g^2 \rangle \ll 1$ we look at the entire polymer, whereas in the limit $x \gg 1$ or $q^2 \langle R_g^2 \rangle \gg 1$ we look at the Gaussian structure inside.

But let us return to our original motivation, the experimental confirmation of our calculation of C_N. We introduce the characteristic ratio $C_N \approx C_\infty$ into (2.104) realising that $b^2 N$ is just $\langle R_N^2 \rangle$. Using the latter's generalization (2.8) (with $\nu = 1/2$), we now consider b a bond length and x becomes $x = q^2 C_\infty N/6$. In addition, we also use b as the unit of length, i.e. q is now in units of $1/b$. Consequently, a plot of $q^2 N P_i(\theta)$ should yield a plateau of the data, i.e. $q^2 N P_i(\theta) \rightarrow 12/C_\infty$ for large q (note: 'large q', as we shall see shortly, really means 'intermediate q' for real

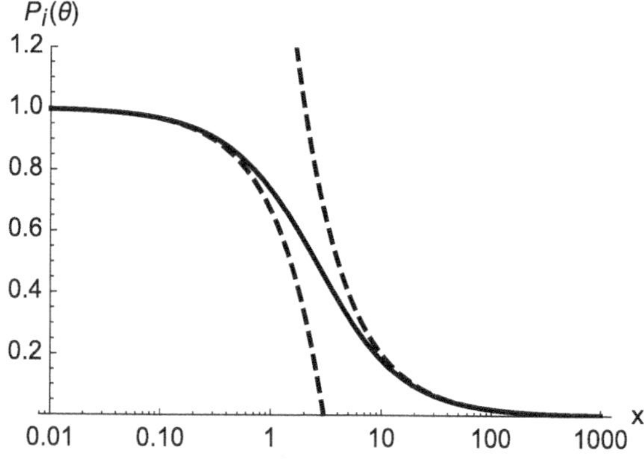

Fig. 2.22 $P_i(\theta)$) versus x according to (2.104). The dashed lines show the two limiting forms

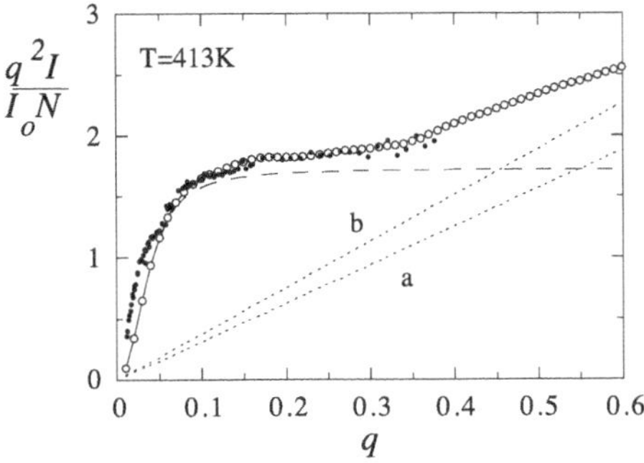

Fig. 2.23 Kratky plot of the reduced scattering intensity obtained from polyethylene in the melt by small-angle neutron scattering [8]. Here $I/(I_o N) = N P_i(\theta)$ and $q = q_{exp} b$ with $b = 1.53$ Å. The data are the solid dots. The long-dashed line shows (2.104) using $C_\infty \approx 7$. The open symbols were obtained using the RIS/TM method to build chains as explained in the text. The two straight dotted lines, labeled a and b, correspond to straight rows of C-atoms spaced 1 unit (**a**) or $\cos 34°$ ≈ 0.83 units (**b**) apart

chains). Such experimental results are shown in Fig. 2.23. Here the reduced intensity is multiplied by q^2 in order to bring out the 'Gaussian plateau', which is called **Kratky plot**. Note that (2.104) is a good approximation to the experimental data and $C_\infty \approx 7$ is in good accord with our result obtain from the RIS/TM method at this temperature.

Figure 2.23 also contains open circles, which appear to be close to the actual data—closer than (2.104). Where do the open symbols come from? The answer brings us back to (2.100) and to the average over chain conformations. Suppose we generate many independent conformations of 'united atom' PE chains by some method—we explain the method below. We then compile a histogram of distances r_{ij} between pairs of united-atom carbon atoms (scattering centers) within each chain and average over all histograms. The result is $n(r, \Delta r)$ shown in Fig. 2.24. Here Δr is the bin width and r is the distance between pairs corresponding to this binning. Note the numerous 'spikes' at small r. There must be, for instance, a pronounced spike at $r = 1$, because there are $N - 1$ pairs separated by exactly one bond distance. In addition, the PE chain is not really a random walk at small C-C separations and these C-C separations appear as discrete spikes.

Using $n(r, \Delta r)$, the scattering intensity becomes

$$P_i(\theta) \approx \frac{1}{N^2} \sum_k n(\Delta r\, k, \delta r) \frac{\sin(q \Delta r\, k)}{q \Delta r\, k}. \tag{2.105}$$

Here the k-summation is over all bins and $r = \Delta r\, k$. We also omit a term $\propto N$.

In order to better understand what is happening for $q > 0.35$, we must look at the next figure, Fig. 2.25, which shows Kratky plots for PE chains of very different lengths. All curves apparently converge when q becomes large. This is because at large q we are probing short distances in real space and on short distances a short

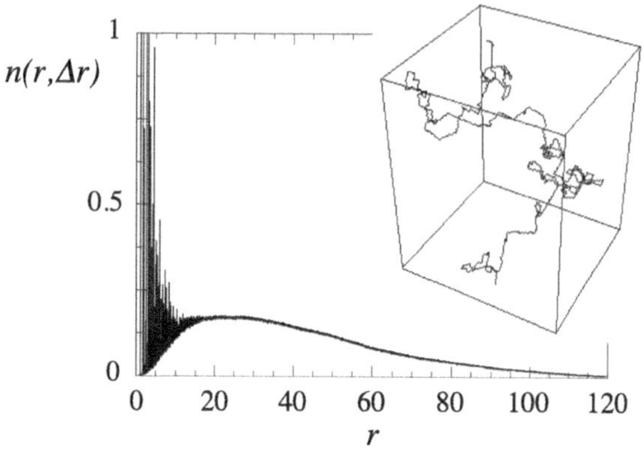

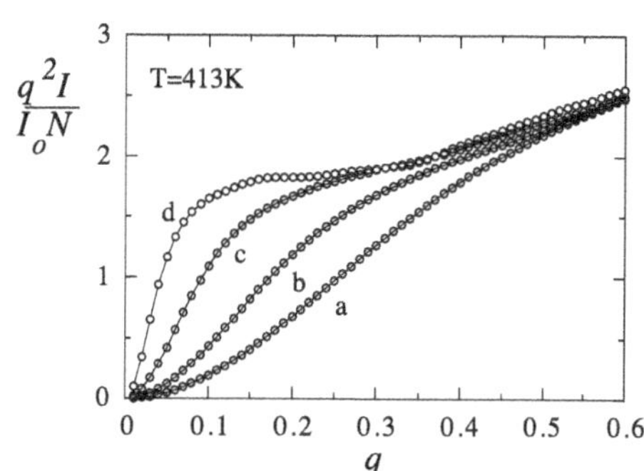

Fig. 2.24 Distance distribution $n(r, \Delta r)$ (shown here truncated at 1) compiled from hundred PE chain conformations at $T = 413$ K using $\Delta r = 0.01$ (in units of the C-C bond length) and $N = 1000$. The inset depicts a selected conformation

Fig. 2.25 Kratky plot of the reduced scattering intensities (note: $I/(I_oN) = N P_i(\theta)$) for numerically generated chains of different length: $N = 20$ (**a**), $N = 50$ (**b**), $N = 200$ (**c**), and $N = 1000$ (**d**)

chain differs little from a long chain. When q is decreased, however, we begin to probe larger and larger distances in real space and a $N = 20$-chain then differs greatly from a $N = 1000$-chain. For PE to be well represented by a Gaussian random walk, i.e. by (2.101), we need chains containing at least several hundred effective C-atoms. By the way, the two straight dotted lines labeled a and b in Fig. 2.23 represent Kratky plots of straight rows of C-atoms spaced 1 (a) unit or $\cos 34°$ (b) units apart.

What is the construction method for the chain conformations used to generate the open symbols-results in Figs. 2.23 and 2.25? Since we want to generate PE chains, we use $\phi = 112°$. The torsion angles are ϑ_{g_-}, ϑ_t, and ϑ_{g_+}, which we had used when we discussed the transfer matrix-method. We also use the potential energy functions (2.18) and (2.42). The bond length is the unit of length. Finally, $P(l \mid k) \equiv P(lk)/P(k)$ is the conditional probability that torsion angle l is the successor of torsion angle k.

For any number of torsion angles the algorithm is the following:

1. Generate a uniform random number z in the interval $[0, 1]$.
2. Continue an existing chain, whose last torsion angle is $\vartheta^{(k)}$, with the new torsion angle

$$\begin{aligned}
\vartheta^{(1)} \text{ if } & & z \le P(1 \mid k) \\
\vartheta^{(2)} \text{ " } & & P(1 \mid k) < z \le P(1 \mid k) + P(2 \mid k) \\
\vartheta^{(3)} \text{ " } & P(1 \mid k) + P(2 \mid k) < z \le P(1 \mid k) + P(2 \mid k) + P(3 \mid k) \\
\vdots \quad \vdots & & \vdots \\
\vartheta^{(m)} \text{ if } & & P(1 \mid k) + \cdots + P(m-1 \mid k) < z.
\end{aligned}$$

The new bond vector $\vec{b}_{i+1}$ is then generated via $\vec{e}_{i+1} = -\vec{e}_i \cos\phi + \vec{e}_i \times \vec{e}_{i-1} \sin\vartheta^{(l)} - (\vec{e}_i \times \vec{e}_{i-1}) \times \vec{e}_i \cos\vartheta^{(l)}$, based on its predecessors. The $\vec{e}$ are unit vectors along the bonds indicated by the subscripts. [Start the chain from two arbitrary but not parallel vectors, e.g., $\vec{e}_{-1} = (1, 0, 0)$ and $\vec{e}_0 = (0, 1, 0)$].

3. Continue with step 1 while $i + 1$ is less than the desired chain length.

In the present case we can use the probabilities according to (2.41) instead of the full conditional probabilities.

Remark: Perhaps you are puzzled by the following question regarding the actual experiment. If the sample is a dense melt, how can we be sure that the two scattering centers i and j do belong to the same chain? The answer is that the experimentalists use a mixture of protonated and deuterated polymers. A small amount of chains in which deuterium replaces the ordinary hydrogen is mixed into a much larger amount of ordinary chains. The diluted and thus isolated deuterated polymers provide a much stronger contrast and 'outshine' the protonated polymers. This does not work with X-rays (or visible light) since the latter interact with the electron shells.

Static Light Scattering Revisited:

This is a good point to return to our discussion of light scattering in Chap. 1 and in particular to (1.18). Equation (1.18) can be combined with (2.104) and a third ingredient into a new equation, which is used to obtain the weight-average molar polymer mass, the radius of gyration, and the second osmotic virial coefficient in a widely used standard procedure called the **Zimm plot** [9]. We first write down this equation, then we provide a justification for it, and finally we discuss its use.

The equation is

$$\frac{K \langle c_w \rangle}{R_\theta} = \left(1 + \frac{1}{3} q^2 \langle R_g^2 \rangle\right)\left(\frac{1}{\bar{M}_w} + 2A_2 \langle c_w \rangle\right). \tag{2.106}$$

The first term in brackets on the right hand side results if we approximate $P_i(\theta)$ by Guinier's law from (2.104). The sign is different because we use $(1 - z)^{-1} \approx 1 + z$ for small z. The second factor, i.e. $(1/\bar{M}_w + 2A_2 \langle c_w \rangle)$, requires additional work however.

In order to find this factor we must look at $P_i(\theta)$ in its form in (2.98), i.e.

$$P_i(\theta) = \frac{1}{\langle v_1 \rangle^2} \frac{1}{V^2} \int_V d^3r \, d^3r' e^{i\vec{q}\cdot(\vec{r}-\vec{r}')} \langle v_1(\vec{r}) v_1(\vec{r}') \rangle, \tag{2.107}$$

where we use $N = m_1 \nu_1$. m_1 is the number of monomers within a polymer and ν_1 is the number of polymers. You may object to this because we just stated 'The first term in brackets on the right hand side results if we approximate $P_i(\theta)$ by Guinier's law from (2.104)', which would mean that we have taken care of $P_i(\theta)$ already. This is not quite right though. What we have done thus far means that we have included the scattering from single chains. However, if we look at our system on a much larger scale we begin to include more and more polymers. The situation is comparable to what we did when we discussed the form factor of a monomer. Guinier's law here essentially is the form factor of a polymer. Now we are interested in the contribution to the scattering intensity on this much larger scale—in fact we are looking at an infinite volume in which the polymer mass density nevertheless is $\langle c_w \rangle$ on average. This means we want to compute $P_i(\theta)$ in the limit $\vec{q} \to 0$. In this limit (2.107) becomes

$$P_{i,f}(0) = \frac{1}{\langle \nu_1 \rangle^2} \frac{1}{V^2} \int_V d^3r d^3r' \langle \nu_1(\vec{r})\nu_1(\vec{r}\,') \rangle. \tag{2.108}$$

Note in particular that we are dealing with an open system in which the number of polymers ν_1 fluctuates. ν_1 is given by

$$\nu_1 = \frac{1}{V} \int_V d^3r \, \nu_1(\vec{r}). \tag{2.109}$$

Hence

$$P_{i,f}(0) = \frac{\langle \nu_1^2 \rangle}{\langle \nu_1 \rangle^2}. \tag{2.110}$$

In standard textbooks on Statistical Mechanics or Statistical Thermodynamics (e.g., [10]; Sect. 5.3.2) the quantity $\langle \nu_1^2 \rangle$ is discussed in the context of the **grand canonical ensemble** applied to a one-component system in which the particle number can fluctuate. Here our system is a solute in a solvent but the general computation remains the same—except that the chemical potential becomes the chemical potential of the solute and the pressure becomes the **osmotic pressure**, i.e.

$$\langle \nu_1^2 \rangle = \frac{\partial \langle \nu_1 \rangle}{\partial (\beta \mu_1)} \bigg|_{\beta,V} = \frac{1}{-\beta \frac{V^2}{\langle \nu_1 \rangle^2} \frac{\partial \Pi_1}{\partial V} \big|_{\beta,\langle \nu_i \rangle}}, \tag{2.111}$$

where μ_1 is the polymer chemical potential and $\beta = (k_B T)^{-1}$.

In Sect. 3.4 we discuss the osmotic pressure of polymers in solution. We shall see that

$$\Pi_1 = \frac{1}{\beta} \left[\frac{\langle \nu_1 \rangle}{V} + B_2 \left(\frac{\langle \nu_1 \rangle}{V} \right)^2 + \mathcal{O}\left(\left(\frac{\langle \nu_1 \rangle}{V} \right)^3 \right) \right] \tag{2.112}$$

in the limit of small polymer concentration and therefore

$$\langle v_1^2 \rangle = (1 + 2B_2 \langle c_w \rangle / \bar{M}_w)^{-1} \langle v_1 \rangle. \tag{2.113}$$

Hence,

$$P_{i,f}(0) = (1 + 2B_2 \langle c_w \rangle / \bar{M}_w)^{-1} \langle v_1 \rangle^{-1} \tag{2.114}$$

(note: $A_2 = B_2/(\bar{M}_w)^2$). The final step is the approximation

$$P_i(\theta) \approx P_{i,G}(\theta) P_{i,f}(0), \tag{2.115}$$

which means that we approximate the total $P_i(\theta)$ by the Guinier law, $P_{i,G}(\theta)$, multiplied with $P_{i,f}(0)$ due to the fluctuation of the number of polymers in the volume. In other words, we assume that the shape of the scattering intensity is well described by Guinier's law and the leading q-independent correction contributed by the polymer-polymer interaction expressed in terms of the **second osmotic virial coefficient**. Figure 2.26 shows the light scattering intensity obtained from a linear

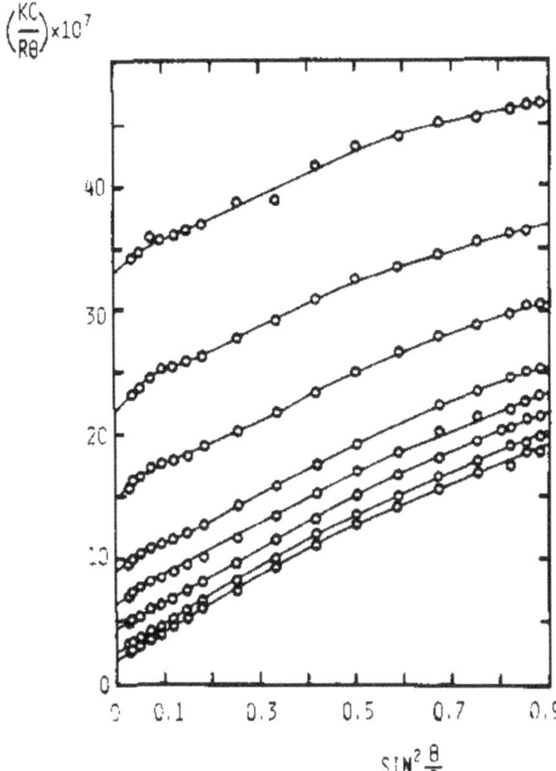

Fig. 2.26 Light scattering intensity from poly(α-methylstyrene) in toluene at 25 °C. Reprinted with permission from [11]. Copyright 1981 American Chemical Society

polymer in solution at variable concentrations (including 0.033, 0.052, 0.097, 0.142, 0.194, 0.288, 0.391, and 0.530 g/dL from bottom to top). Note that the weight-average molecular weight of the polymer is $7.47 \cdot 10^6$ g/mol. The data are reasonably linear, and therefore in line with our 'ad hoc approximation (2.115), at the lowest polymer concentrations only.

Let's discuss how to use (2.106). For instance, we can try to plot $\frac{K\langle c_w \rangle}{R_\theta}$ versus $\langle c_w \rangle$ at $q = 0$. From the intercept of a line through the data with the y-axis we would find $1/\bar{M}_w$ and the slope gives us $2A_2$. However, data at $q = 0$ are not available, since this means looking directly into the incoming beam. Likewise we cannot plot $\frac{K\langle c_w \rangle}{R_\theta}$ versus q^2 at $\langle c_w \rangle = 0$ to obtain $1/\bar{M}_w$ from the intercept and $\langle R_g^2 \rangle/(3\bar{M}_w)$ from the slope. Without polymer we cannot make this measurement. Instead, measurements are carried out which yield $\frac{K\langle c_w \rangle}{R_\theta}$ for a series of non-zero concentrations and non-zero scattering angles as shown in the Zimm plot in Fig. 2.27. The trick is that on the x-axis it is not just q^2 but $q^2 + c_o\langle c_w \rangle$ which is plotted, where c_o is an arbitrary constant. This shift of q^2 will not change the slope or intercept of the $c_w = 0$-line, which we construct by drawing a line parallel to the thin blue lines through the data for $\langle c_{w,1} \rangle$, $\langle c_{w,2} \rangle$, $c_{w,3} \rangle$, and $\langle c_{w,4} \rangle$. However, this line is subject to a constraint. The constraint is that the line, the thick blue line in Fig. 2.27, must, on the y-axis, intersect with another line, the thick red line, drawn parallel to the red lines through the data for q_1, q_2, and q_3. This is because for $\langle c_w \rangle = q = 0$ we must have $\frac{K\langle c_w \rangle}{R_\theta} = 1/\bar{M}_w$ (independent of c_o!). However, the aforementioned constraint by itself is not sufficient. We need to find at least one of the thick lines, i.e. the thick blue or the thick red line, via extrapolation based on the spacings of the corresponding thin lines, i.e. the thin blue or the thin red lines. Ideally we carry out the extrapolation for both and find that the resulting thick lines intersect on the y-axis. Hence, from a single plot, the Zimm plot, we are able to extract $\bar{M}_w$, $\langle R_g^2 \rangle$, as well as B_2. Problem 2.7 lets you practice how to do this (for $\bar{M}_w$).

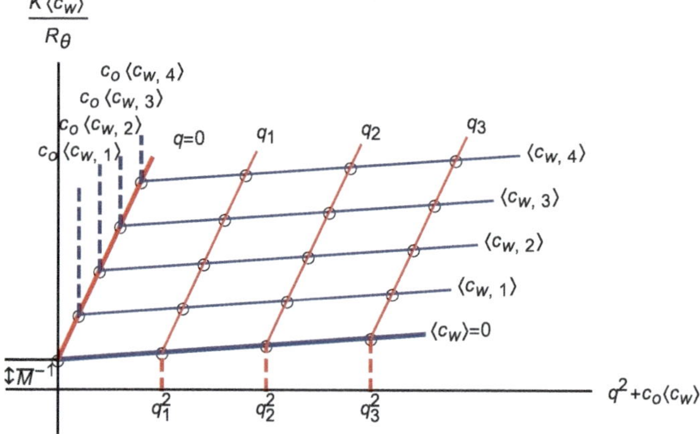

Fig. 2.27 Illustration of a Zimm plot

Remark 1: The procedure also works if fitting the data requires slightly bent lines!

Remark 2: An expression for B_2 in (2.106) (note again: $A_2 = B_2/(\bar{M}_w)^2$) is given in (3.40) in the context of osmotic pressure in a dilute polymer in a solvent, computed using a mean field lattice model.

Scattering from Non-ideal Chains:

Thus far we have sidestepped the exponent ν. What is the effect of $\nu \neq 1/2$ on the scattering intensity? Or in other words, how can we determine ν from scattering experiments?

The answer to this question may require a new $p(R)$, which means a more general expression for the probability that the ends of a polymer chain are separated by the distance R. When the chains are ideal, i.e. the chains can intersect and $\nu = 1/2$, $p(R)$ is given by (2.67). In the case of general ν, which includes intersecting chains, (2.81) can be used to deduce the attendant $p(R)$. Note that (2.81) is a configuration free energy equal to $-TS = -Tk_B \ln p(R) + const$. This immediately tells us that

$$p(R) \propto \exp[-k_\nu (R/bN^\nu)^{1/(1-\nu)}], \tag{2.116}$$

where k_ν is a factor equal to $3/2$ if $\nu = 1/2$. In principle we can use this $p(R)$ to express $p(r_{ij})$ in (2.100) and then proceed as before with the calculation of the scattering intensity. However, this is quite tedious and another approach is better and provides greater physical insight.

Before we start, let us briefly look at the effect of ν on $\langle R_g^2 \rangle$. We make a short calculation, i.e.

$$\sum_{i>j}^{N} |i-j|^{2\nu} = (N-1)1^{2\nu} + (N-2)2^{2\nu} + \cdots$$

$$\approx \int_1^N dx (N-x)x^{2\nu} \approx \frac{N^{2\nu}}{(2\nu+1)(2\nu+2)}$$

for large N. Hence, in this limit, we have

$$\langle R_g^2 \rangle \propto N^{2\nu}. \tag{2.117}$$

In other words, here it does not really matter whether we look at a linear polymer or maybe at a polymer with a more complex architecture.

What we have is the general relation

$$n \sim r^{1/\nu}. \tag{2.118}$$

The quantity n is the number of monomers or Kuhn segments, which we expect to find inside a circle with radius r. Ordinarily we expect the number of things to increase with the dth power in a d-dimensional uniform system. In a poor solvent in three dimensions, for instance, in which the polymer essentially collapses into a dense drop, we have $1/\nu = 3$ (cf. Table 2.4). Here, however, we usually look at systems possessing regions, which come in all sizes, devoid of monomers or Kuhn segments. We can tell because generally $1/\nu < d$. Hence, we are looking at **fractals** whose **(mass) fractal dimension** is

$$d_f \equiv 1/\nu. \tag{2.119}$$

So, how do we calculate the scattering intensity of a fractal structure based on (2.118)? We can use (2.105). Remember that $n(\Delta r k, \Delta r)$ is the number of scatterers in a shell of thickness Δr and radius r. Based on (2.118) this means that

$$n(r, dr) \sim r^{d_f - 1}, \tag{2.120}$$

where we have made the transition to the infinitesimal form, $\Delta r \to dr$ and $\Delta r\, k \to r$. Consequently the summation of k is replaced by an integration over r, i.e. $\sum_k \to \int dr$. Hence,

$$I(q) \sim \int_0^\infty dr\, r^{d_f - 1} e^{-r/\xi} \frac{\sin(qr)}{qr}. \tag{2.121}$$

The exponential is a convenient cutoff, accounting for the finiteness of the fractal. Using $x = qr$ this becomes

$$I(q) \sim \frac{1}{q^{d_f}} \int_0^\infty dx\, x^{d_f - 1} e^{-x/(q\xi)} \frac{\sin x}{x} \sim q^{-d_f} \text{ if } q \gg \xi^{-1}. \tag{2.122}$$

The integral is a slowly varying function of $q\xi$ if this quantity is significantly larger than unity, i.e. this condition means that we are looking inside the fractal objects. Clearly, there is another limit if q becomes 'too large'. 'Too large' here means that we begin to see the details within the fractal which themselves possess a different structure. But in the intermediate regime $I \sim q^{-d_f}$ or in the case of polymers

$$I(q) \sim q^{-1/\nu} \tag{2.123}$$

holds.

If we apply this formula to the ideal chain or $\nu = 1/2$ we find $q^2 I(q) \sim const$, which is the plateau of the dashed line in Fig. 2.23. Non-ideal chains in three dimensions yield $q^2 I(q) \sim q^{1/3}$ instead. Note that the experimental data in this figure indeed exhibit a slow increase. But this increase is also obtained with the transfer matrix model, which produces ideal chains. Thus, in this case, the slow increase in the 'plateau regime' is likely due to the steeper increase at still larger q, where the observed objects do not possess the expected fractal structure. In addition, other

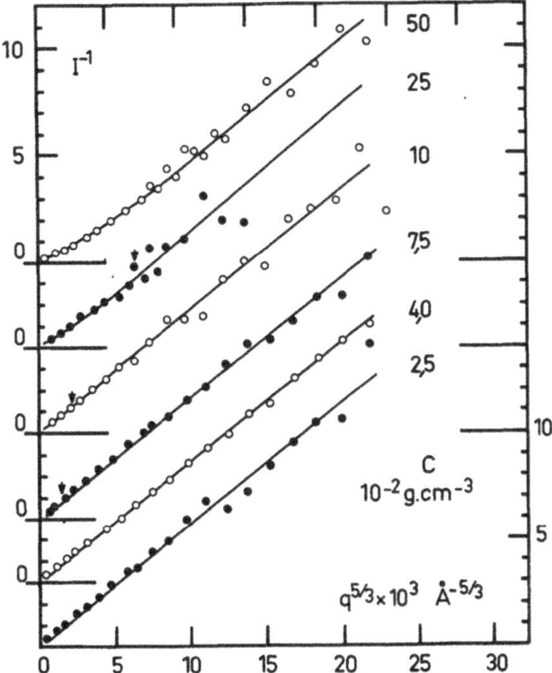

Fig. 2.28 Inverse scattering intensity versus momentum transfer raised to the power 5/3. Reprinted with permission from [17]. The polymer is polystyrene dissolved in carbon disulfide. The data sets are results for a fixed concentration of deuterated PS $(0.005\,\mathrm{g\,cm^{-3}})$ within an overall variable amount of protonated PS

experiments have confirmed that chains in melts are indeed ideal chains (e.g., [12–14]). The general idea is to carry out experiments in a sufficiently wide q-range, either on melts or polymer solutions, and then plot $\ln I(q)$ versus $\ln q$. This should yield a straight line with slope $-1/v$ in the proper q-range. Also possible are plots where the inverse intensity is plotted versus $q^{5/3}$ as in Fig. 2.28. This is an example of a study where the authors look at the cross-over between $v = 1/2$ versus $v = 3/5$ in solutions at variable polymer concentration.

Overlap Concentration:

A concentration of particular importance is the **overlap concentration** ρ^* at which the volumes occupied by individual polymers begin to overlap. We expect this to happen when the overall monomer concentration is the same as the monomer concentration within the volume occupied by the isolated polymer, i.e.

$$\rho^* = \frac{N}{R_g^d} \sim \frac{N}{N^{dv}} = N^{1-dv}. \tag{2.124}$$

Here $R_g \equiv = \sqrt{\langle \vec{R}_g^2 \rangle}$. For $d = 3$ this means $\rho^* \sim N^{-1/2}$ when the chains are ideal and $\rho^* \sim N^{-4/5}$ when they are not. Note that ρ^* is very small when the chains are long.

Question: What is ρ^* in the system for which light scattering data are shown in Fig. 2.26? Assume 'good solvent'. You should find that ρ^* is less or at best about equal to the lowest of the concentrations in that figure.

However, we want to move on and not get overwhelmed by what can be done and a multitude of power laws. Additional information on the topic of scattering from polymers can be found in a book edited by Glatter and Kratky [15]—in particular in the article by Kirste and Oberthur (Synthetic Polymers in Solution) or in [16].

2.6 Problems

Problem 2.1   (a) Show that

$$\sum_{i>j} \langle \mathbf{t} \rangle^{i-j} = N \langle \mathbf{t} \rangle \frac{1}{\mathbf{I} - \langle \mathbf{t} \rangle} - \langle \mathbf{t} \rangle \left(\mathbf{I} - \langle \mathbf{t} \rangle^N \right) \left(\frac{1}{\mathbf{I} - \langle \mathbf{t} \rangle} \right)^2. \tag{2.125}$$

(b) Write down an expression for C_N for all N.

(c) Using the result obtained in part (b) show that the mean square end-to-end distance for an ideal persistent flexible polymer is given by

$$\langle R^2 \rangle = 2L P - 2P^2(1 - e^{-L/P}), \tag{2.126}$$

where L is the polymer's contour length and P is its persistence length.
(d) Compare the formula in part (c) to (2.27).

Problem 2.2   (a) We practice the application of the transfer matrix method assuming a very simple polymer model. Every monomer has a potential energy equal to zero. However, adjacent monomers couple via the interaction potential energy $U(\vartheta, \vartheta') = -k_B J \vartheta \vartheta'$, where J is a positive constant (in units of Kelvin). Here we consider a closed loop in the limit of large N. The angular variables can be either -1 or $+1$. Write down the 2×2-transfer matrix and obtain its eigenvalues.

(b) Calculate the internal energy and the heat capacity of the model. In addition calculate the probability $p_T(\vartheta = 1)$.

Problem 2.3   The sketch in Fig. 2.29 shows a path, i.e. a model polymer, on a square lattice (2D) composed of the two elements 1 and 2. Element 1 has the energy $\epsilon_1 = 0$ and element 2 has the energy $\epsilon_2 = \epsilon > 0$. This means that straight segments are energetically favored compared to kinks. There are no interactions between the segments.

Fig. 2.29 Polymer construction on a square lattice

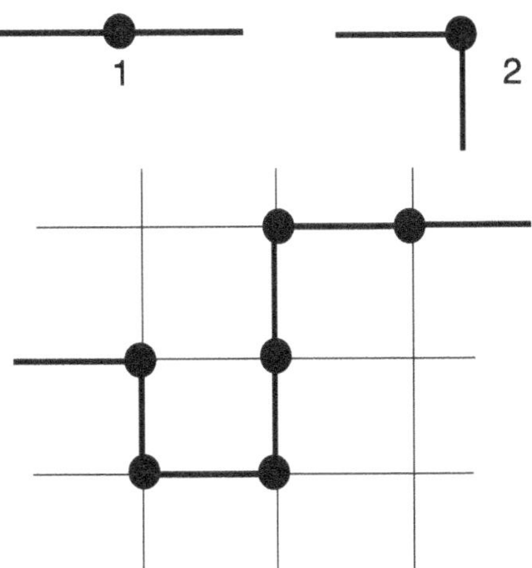

(a) Calculate the partition function for a polymer of length n (segments). Note that there are two kinks (right and left) of type 2.

(b) Calculate the internal energy of the polymer chain.

(c) Calculate the entropy of the polymer chain in the limit $T \to \infty$.

Problem 2.4 We consider the conformation free energy ΔF in the context of the self-consistent field method (cf. (2.50)).

(a) Derive the following integral form of ΔF:

$$\frac{\Delta F}{nk_B T} = \int dx \left[\frac{a^2}{2} \left(\frac{d\psi_0}{dx} \right)^2 + \beta u(x)\psi_0^2(x) \right]. \tag{2.127}$$

Use the constraint

$$1 = \int dx\, \psi_0^2(x) = \int dx\, c(x). \tag{2.128}$$

Hint: $\Delta F = \Delta E - T\Delta S$, where ΔE is expressed as an integral involving $c(x)$ and $u(x)$.

(b) Via the **Lagrange multiplier** method, show that (2.51) follows from (2.127) and (2.128).

Problem 2.5 The central limit theorem states that the random variable

Table 2.6 Selected light scattering data from Fig. 2.26

$\sin^2\left(\frac{\theta}{2}\right)$	$10^{-14} q^2$			$10^7 \frac{K\langle c_w\rangle}{R_\theta}$		
0.09	1.67	4.04	4.72	6.41	8.64	11.39
0.18	3.35	6.13	6.85	8.29	10.27	12.81
0.25	4.65	7.55	8.47	9.83	11.81	14.43
0.33	6.14	9.43	10.14	11.67	13.57	16.02

$$S_n = \frac{\sum_{i=1}^{n} s_i - n\langle s\rangle}{\sqrt{n}\sigma_s} \tag{2.129}$$

possesses a standard normal distribution in the limit $n \to \infty$. Assume that the s_i are uniformly distributed random variables on the interval $(0,1)$. Generate properly normalized S_n-histograms on the computer for $n = 1$ and $n = 5$ and compare them graphically to the standard normal distribution $f(S) = (2\pi)^{-1/2} \exp[-S^2/2]$.

Problem 2.6　　There is a detail regarding the comparison of (2.88) with (2.59) which should not escape our attention. By starting from (2.87) we assume that the condition $bN^\nu \gg D$ is satisfied, or $bN^{1/2} \gg D$ in the case of the ideal chain. In the opposite limit the chain is not constraint by the tube and we cannot expect the above results for $F(D)$ to hold. Since we successfully compared $F^{(}D)$ with the self-consistent field result (2.59), we should wonder where the condition $bN^{1/2} \gg D$ is hidden in the self-consistent field calculation!

Problem 2.7　　We want to practice using the Zimm plot illustrated in Fig. 2.27. Table 2.6 contains selected light scattering data from Fig. 2.26. The concentrations corresponding to the $\frac{K\langle c_w\rangle}{R_\theta}$-values in the table are 0.033, 0.052, 0.097, 0.142, and 0.194 g/dL from left to right. Here the wavelength of the incident light is 436 nm. Since the medium is toluene, this number must be divided by the refractive index of toluene, i.e. $n = 1.496$, to obtain the wavelength with which to calculate q via (2.91). Make a Zimm plot of the data and obtain $\bar{M}_w$. Compare your result with the value for the α-111 sample in [11]—the original source of the measurements.

References

1. W.L. Jorgensen, J.D. Madura, C.J. Swenson, Optimized intermolecular potential functions for liquid hydrocarbons. J. Am. Chem. Soc. **106**, 6638 (1984)
2. T. Odijk, Theory of lyotropic polymer liquid crystals. Macromolecules **18**, 2313 (1986)
3. R. Hentschke, *Statistische Mechanik* (Wiley-VCH, Weinheim, 2004)
4. P.J. Flory, *Statistical Mechanics of Chain Molecules* (Interscience, New York, 1969)
5. H.A. Kramers, G.H. Wannier, Statistics of the two-dimensional ferromagnet. Part II. Phys. Rev. **60**, 263 (1941)
6. A. Pelissetto, E. Vicari, Critical phenomena and renormalization-group theory. Phys. Rept. **368**, 549 (2002)

7. S.-T. Sun, I. Nishio, G. Swislow, T. Tanaka, The Coil-Globule transition: radius of gyration of polystyrene in cyclohexane. J. Chem. Phys. **73**, 5971 (1980)
8. G. Lieser, E.W. Fischer, K. Ibel, Conformation of polyethylene molecules in the melt as revealed by small-angle neutron scattering. J. Polym. Sci. Polym. Lett. **13**, 39 (1975)
9. B.H. Zimm, Apparatus and methods for measurement and interpretation of the angular variation of light scattering; preliminary results on polystyrene solutions. J. Chem. Phys. **16**, 1099 (1948)
10. R. Hentschke, *Thermodynamics*, 2nd edn. (Springer, Heidelberg, 2022)
11. I. Noda, N. Kato, T. Kitano, M. Nagasawa, Thermodynamic properties of moderately concentrated solutions of linear polymers. Macromolecules **14**, 668 (1981)
12. D.G.H. Ballard, G.D. Wignall, J. Schelten, Measurement of molecular dimensions of polystyrene chains in the bulk polymer by low angle neutron diffraction. Eur. Polymer J. **9**, 965 (1973)
13. J.P. Cotton, D. Decker, H. Benoit, B. Farnoux, J. Higgins, G. Jannink, R. Ober, C. Picot, J. des Cloizeaux, Conformation of polymer chain in the bulk. Macromolecules **7**, 863 (1974)
14. R.G. Kirste, W.A. Kruse, K. Ibel, Determination of the conformation of polymers in the amorphous solid state and in concentrated solution by neutron diffraction. Polymer **16**, 120 (1975)
15. O. Glatter, O. Kratky, *Small Angle X-ray Scattering* (Academic, New York, 1982)
16. G. Strobl, *The Physics of Polymers*, 2nd edn. (Springer, Berlin, 1997)
17. B. Farnoux, F. Boue, J.P. Cotton, M. Daoud, G. Jannink, M. Nierlich, P.-G. de Gennes, Crossover in polymer solutions. J. Phys. **39**, 77 (1978)

Chapter 3
Thermodynamics of Blends, Solutions, and Networks

Abstract This section, at first glance, may appear unrelated to our exposition of single chain statistical thermodynamics. This is because the focus is no longer on conformations of single chains or quantities like the end-to-end distance or the radius of gyration. Instead we focus on the combinatorial or **packing entropy** of many chains in a certain volume as well as on their free enthalpy. In the following we develop a simple mean field theory on the lattice, which nevertheless can be applied to blends, solutions, and networks. There are cases when our results will not be correct and will have to be revised. However, the lattice approach is certainly worthwhile, since it provides an overview without immediately blinding us with complexity.

3.1 A Lattice Model for Polymer Mixtures and Solutions

In the following we study binary mixtures assuming that the two components are linear polymers. We know already that a simple but instructive approximation of a linear polymer is a path on a lattice as depicted in Fig. 3.1. Note that the polymers in this figure are quite similar to the polymer depicted in Fig. 2.16. Except that here we do not focus on the end-to-end distance. The lattice is a square lattice and every lattice cell contains one polymer segment or Kuhn segment. Kuhn segments belonging to the same polymer are connected by a solid line. The solid and hollow circles indicate two chemically different types of Kuhn segments. In the following we consider v_i polymers of type i with length (or 'mass') m_i ($i = 1, 2$). This means that all polymers of type i possess the same length, i.e. they are **monodisperse**. In reality (technical) polymers are **polydisperse**, i.e. they do have a distribution of lengths. Here we avoid this complication. There are $N = N_1 + N_2$ Kuhn segments total ($N_i = m_i v_i$) and N is equal to the number of lattice cells (Here and in the following we use N or N_i to count Kuhn segments, because n or n_i is reserved for the number of moles of a species.). This means that the lattice is fully occupied.

Having specified our model we want to estimate the number of distinct polymer configurations on the lattice:

R. Hentschke, *A Concise Introduction to Polymer Physics*, Undergraduate Lecture Notes in Physics, https://doi.org/10.1007/978-3-031-87324-9_3

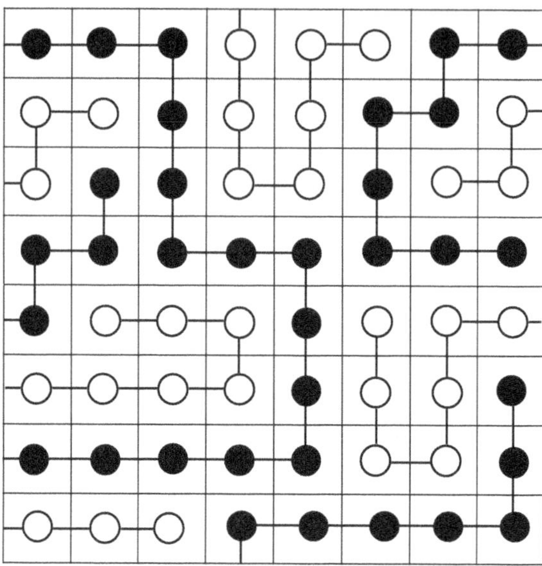

Fig. 3.1 Binary mixture of linear polymers represented by paths on a lattice

(i)   We proceed by severing all bonds connecting the Kuhn segments in each polymer chain. The individual Kuhn segments, which we consider distinguishable at this point, are then placed on an empty but otherwise identical lattice. There are $N!$ ways to accomplish this.

(ii)   Now we ask: What is the probability that in one such configuration all Kuhn segments do have the same neighbors which they had before?

We first approximate the probability that a particular Kuhn segment is placed in a cell next to its polymer-neighbor Kuhn segment via

$$\left(\frac{q-1}{N}\right)^{m_1-1} \quad \text{or} \quad \left(\frac{q-1}{N}\right)^{m_2-1}.$$

Here q is the **coordination number** of the lattice. This is the number of neighbors each cell has. On a square lattice $q = 4$; on a simple cubic lattice $q = 6$. This means that if we have a polymer partially laid out on the lattice and we place the next Kuhn segment, of which we know that it is the neighbor in the polymer, on the lattice blindfolded, then there are $q - 1$ 'good' cells compared to N cells total. Of course we neglect occupancy of the cells by previous Kuhn segments—a truly crude approximation. Nevertheless, we approximate the above probability by

$$\left(\frac{q-1}{N}\right)^{(m_1-1)\nu_1} \left(\frac{q-1}{N}\right)^{(m_2-1)\nu_2}.$$

(iii)   This number we multiply by $N!$, the total number of configurations. But we also must divide by the product $v_1!v_2!$, because two polymers of the same type are indistinguishable. All in all we find that the number of distinguishable ways to accommodate the polymers on the lattice, Ω, may be approximated via

$$\Omega \approx \frac{N!}{v_1!v_2!} \left(\frac{q-1}{N} \right)^{(m_1-1)v_1} \left(\frac{q-1}{N} \right)^{(m_2-1)v_2} . \tag{3.1}$$

We can now work out the attendant entropy, which is the configuration entropy, using

$$S = k_B \ln \Omega . \tag{3.2}$$

Employing **Stirling's formula**, i.e.

$$\ln N! \approx N \ln N - N + \ln \sqrt{2\pi N} \approx N \ln N - N \quad \text{(if N is large)} \tag{3.3}$$

(see for instance Chap. 6 in [1]), we obtain

$$\frac{S}{Nk_B} = -\frac{\phi_1}{m_1} \ln \frac{\phi_1}{m_1} - \frac{\phi_2}{m_2} \ln \frac{\phi_2}{m_2} \tag{3.4}$$
$$+ \left[(1 - \frac{1}{m_1})\phi_1 + (1 - \frac{1}{m_2})\phi_2 \right] \ln \frac{q-1}{e} ,$$

where $\phi_i = N_i/N$.

Before we discuss this, we compute the **mixing entropy**:

$$\frac{\Delta S}{k_B} = -v_1 \ln \phi_1 - v_2 \ln \phi_2 . \tag{3.5}$$

This is the entropy change when we combine two lattices of size N_1 and N_2, each filled with the respective polymers of type 1 and 2, into one lattice of size $N = N_1 + N_2$, i.e.

$$\Delta S = S - S_1 - S_2 , \tag{3.6}$$

where $S_i = k_B \ln \Omega_i$ and

$$\Omega_i \approx \frac{N_i!}{v_i!} \left(\frac{q-1}{N_i} \right)^{(m_i-1)v_i} . \tag{3.7}$$

Using again $v_i = N_i/m_i$ and $\phi_i = N_i/N$ (3.5) becomes

$$\frac{\Delta S}{Nk_B} = -\frac{\phi_1}{m_1} \ln \phi_1 - \frac{\phi_2}{m_2} \ln \phi_2 . \tag{3.8}$$

Interaction on the Lattice:

Thus far we have only expressions for the configuration entropy and the attendant configuration entropy of mixing. We still need to construct an expression for the interaction free enthalpy in a binary system.

The likelihood for a lattice site to be occupied by a 1- or a 2-polymer is ϕ_1 or ϕ_2, respectively. The number of $1 - 2$-contacts (neighboring segments) is given by

$$\approx \phi_1 \phi_2 N q \; .$$

Analogously we obtain for the average number of $1 - 1$- and $2 - 2$-contacts

$$\approx \frac{1}{2} \phi_1 \phi_1 N q$$

and

$$\approx \frac{1}{2} \phi_2 \phi_2 N q \; ,$$

respectively. The factor $1/2$ prevents overcounting of contacts. Finally, we assign an interaction free enthalpy g_{12}, g_{11}, and g_{22} to each type of contact. Hence the total interaction free enthalpy becomes

$$G_i = \left(g_{12} \phi_1 \phi_2 + \frac{1}{2} g_{11} \phi_1^2 + \frac{1}{2} g_{22} \phi_2^2 \right) N q \; . \tag{3.9}$$

This is not yet the final expression, since it is common to use the following definitions:

$$\chi \equiv -\frac{q}{2 k_B T} \left(g_{11} + g_{22} - 2 g_{12} \right) \tag{3.10}$$

$$\chi_1 \equiv -\frac{q}{2 k_B T} g_{11} \tag{3.11}$$

$$\chi_2 \equiv -\frac{q}{2 k_B T} g_{22} \; . \tag{3.12}$$

This means that (3.9) becomes

$$\frac{G_i}{N k_B T} = \chi \phi_1 \phi_2 - \left(\chi_1 \phi_1 + \chi_2 \phi_2 \right) \phi \; . \tag{3.13}$$

The first term is the interaction free enthalpy of mixing

$$\frac{\Delta G_i}{N k_B T} = \chi \phi_1 \phi_2 \; . \tag{3.14}$$

This agrees with our intuition, because $-g_{11} - g_{22} + 2g_{12}$ means that the formation of two $1-2$-contacts comes at the expense of one $1-1$- and one $2-2$-contact.

The combination of (3.8) and (3.14) yields

$$\Delta g \equiv \frac{\Delta G}{N k_B T} = \frac{\Delta G_i}{N k_B T} - \frac{\Delta S}{N k_B} \tag{3.15}$$

and thus the important equation

$$\Delta g = \frac{\phi_1}{m_1} \ln \phi_1 + \frac{\phi_2}{m_2} \ln \phi_2 + \chi \phi_1 \phi_2 . \tag{3.16}$$

Equation (3.16) is the **Flory–Huggins equation** describing a binary polymer mixture or a polymer in solution. The quantity χ is the **Flory–Huggins parameter** or **χ-parameter**.

Remark: You may wonder why we do not call the g_{ij} simply interaction 'energy' or interaction 'enthalpy'. The point is that we add these expressions to $(-T\times)$ the configuration entropy or the configuration entropy of mixing to obtain the full free enthalpy or free enthalpy of mixing. However, $S = -dG/dT|_{P,n_i}$ and since the g_{ij} themselves may depend on T, if the fit of the theory to the data requires it, this means that the g_{ij} themselves contribute to the entropy.

The lattice approach outlined here was pioneered independently by Staverman and van Santen [2], Huggins [3] and Flory [4] (cf. [5]).

3.2 Phase Separation in Polymer Mixtures

Before we come to the actual topic, let's take a look at two special types of phase separation.

A Digression—One-component Gas-Liquid Phase Behavior:

Equations (3.4) and (3.8) can be applied to a number of interesting situations. We introduce the replacements $\phi_1 = \phi, m_1 = m$, and $\phi_2 = 1 - \phi$. In addition we assume $m_2 = 1$. This corresponds to polymers in a solvent, where the index 2 indicates the solvent. The resulting configuration entropy is

$$\frac{S}{nR} = -\frac{\phi}{m} \ln \frac{\phi}{m} - (1 - \phi) \ln(1 - \phi) + \phi(1 - \frac{1}{m}) \ln \frac{q-1}{e} . \tag{3.17}$$

If we replace the solvent cells by empty cells, we describe the same type of physical situation described by the **van der Waals equation**. Here the total volume is $V = b^3 N$, where b^3, the cell size, also is the monomer size. We may obtain the attendant configurational pressure via

$$P_{conf} = -\frac{\partial}{\partial V}(-TS)\Big|_T = \frac{RT}{N_A b^3}\left[-\phi(1 - \frac{1}{m}) - \ln(1 - \phi)\right] . \tag{3.18}$$

Analogous to the van der Waals approach we must add a term accounting for attractive interaction between the monomers. Our choice, in analogy to the van der Waals equation of state, is

$$\frac{N_A b^3 P}{RT} = \frac{N_A b^3 P_{conf}}{RT} - \frac{1}{2}\frac{N_A \epsilon_o}{RT}\phi^2 . \tag{3.19}$$

Here $\epsilon_o > 0$ is a parameter whose unit is energy.

The closeness of this equation of state to the van der Waals equation of state becomes even more evident if we compute the gas-liquid critical parameters via $\frac{\partial P}{\partial V}\big|_T = \frac{\partial^2 P}{\partial V^2}\big|_T = 0$. We find

$$T_c = \frac{N_A \epsilon_o}{R}\frac{m}{(\sqrt{m} + 1)^2} \tag{3.20}$$

$$\phi_c = \frac{1}{\sqrt{m} + 1} \tag{3.21}$$

$$\frac{b^3 P_c}{\epsilon_o} = \frac{\frac{1}{2} - \sqrt{m}\left(1 + \sqrt{m}\ln\left(\frac{\sqrt{m}}{\sqrt{m}+1}\right)\right)}{\left(\sqrt{m} + 1\right)^2} . \tag{3.22}$$

In the limit $m = 1$ we therefore have

$$\frac{RT_c}{N_A \epsilon_o} = \frac{1}{4} \qquad \phi_c = \frac{1}{2} \qquad \frac{b^3 P_c}{\epsilon_o} = \frac{2\ln 2 - 1}{8} . \tag{3.23}$$

In addition we may work out the relation between the **critical temperature** T_c and the **Boyle temperature**, i.e.

$$T_{Boyle} = 4T_c , \tag{3.24}$$

or the **critical compressibility factor**

$$\frac{N_A P_c}{RT_c \rho_c} = 2\ln 2 - 1 \approx 0.39 . \tag{3.25}$$

Both values are very close to the same quantities in the van der Waals theory.

Fig. 3.2 Critical compressibility factor for n-alkanes

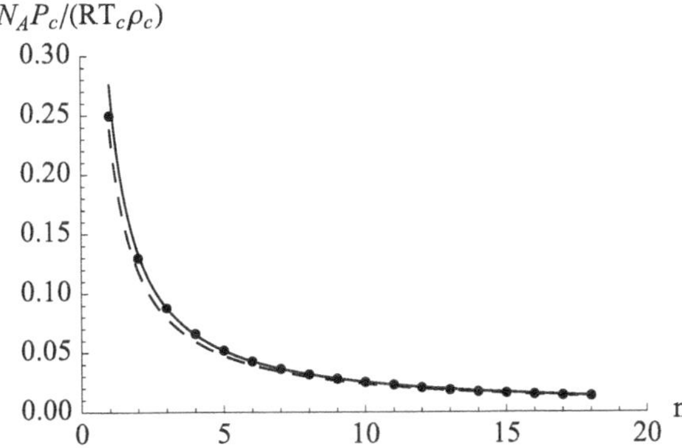

But we do not want to enter into a competition with the van der Waals equation. We therefore focus on the opposite limit, i.e. very long polymer chains, which is not described by the van der Waals equation. In the limit $m \to \infty$ we have to leading order

$$\frac{RT_c}{N_A \epsilon_o} \approx 1 \qquad \phi_c \approx \frac{1}{m^{1/2}} \qquad \frac{b^3 P_c}{\epsilon_o} \approx \frac{1}{3m^{3/2}} \,. \tag{3.26}$$

The corresponding leading behavior of the critical compressibility factor is

$$\frac{N_A P_c}{RT_c \rho_c} \approx \frac{1}{3m} \,. \tag{3.27}$$

Here ρ_c is the number density of the monomer units—not the polymers! Notice that the critical compressibility factor is not a constant independent of the type of molecule as before. Figure 3.2 shows the critical compressibility factor for n-alkanes ($1 \leq n \leq 18$), where the symbols are data from [6]. The mass density is converted to the monomer number density using CH_2 as the monomer unit. This also implies $m = n$. The lines are fits to the data using the full expressions, i.e. (3.20) to (3.22), (solid line) and the limiting law, (3.27), (dashed line). The only fit parameter is a multiplicative constant, i.e. instead of $1/(3m)$ we use $0.24/m$ to match the data for large m.

Note also that expressing pressure, temperature, and volume or density in terms of their critical values eliminates the material parameters b^3 and ϵ_o, but it does not eliminate m. This means that the resulting equation of state is not universal in the sense that it is different for molecules distinguished by their length m. Hence, the **law of corresponding states** is not obeyed by molecules with different m.

> **Another Digression—a Model for Liquid-Liquid Coexistence:**

Another type of phase separation is observed in binary low molecular weight mixtures. Depending on thermodynamic conditions the components may be **miscible** or not. A simple model describing this is based on the following molar free enthalpy approximation:

$$g = x_A^{(l)} g_A + x_B^{(l)} g_B + x_A^{(l)} \ln x_A^{(l)} + x_B^{(l)} \ln x_B^{(l)} + \chi x_A^{(l)} x_B^{(l)} . \tag{3.28}$$

Here g_A and g_B are the (reduced) molar free enthalpies of two pure liquid components A and B. Mixing A and B gives rise to the **mixing free enthalpy** described by the ln-terms. Note that the mole fractions are $x_A^{(l)}$ and $x_B^{(l)} = 1 - x_A^{(l)}$. We immediately see that this description is identical to the lattice model which we have developed, since $\phi_1 = N_1/(N_1 + N_2) = x_1 \equiv x_A^{(l)}$ and $\phi_2 = N_2/(N_1 + N_2) = x_2 \equiv x_B^{(l)}$.

Figure 3.3 shows g for different values of the χ-parameter. If χ is less than a critical value, then g is a convex function of $x_A^{(l)}$ (or $x_B^{(l)}$). This situation is analogous to the free energy in the van der Waals theory for temperatures above the critical temperature. If $\chi = \chi_c$, then the curvature of g at $x_A^{(l)} = 1/2$ becomes zero. For still larger values of χ a hump develops—again analogous to the free energy in the van der Waals theory at temperatures less than the critical temperature. Driven by the second law the system now lowers its free enthalpy by separating into two types of regions, which over time will coagulate into two large domains, one depleted of A and one enriched with A. The resulting **phase diagram** is shown in the lower right panel in Fig. 3.3. The **binodal line** is obtained via a **common tangent construction** applied to the free enthalpy (cf. the lower left panel). Note

$$\left.\frac{\partial g}{\partial x_A}\right|_{T,P} = \frac{1}{k_B T} \left.\frac{\partial G}{\partial N_A}\right|_{T,P} = \frac{\mu_A}{k_B T} . \tag{3.29}$$

This means that the chemical potential of A is the same along the tangent. The common tangent is the lowest possible free enthalpy between $x_{A,poor}$ and $x_{A,rich}$. For a given x_A in this range the quantity $(x_A - x_{A,poor})/(x_{A,rich} - x_{A,poor})$ is the fraction of A in the $A, rich$-phase relative to the total amount of A in the system. The second special line is the **spinodal line**. It indicates the stability limit

$$\left.\frac{\partial^2 g}{\partial x_A^2}\right|_{T,P} = 0 \tag{3.30}$$

(cf. [7]; the third condition in (3.16)). Both lines meet at the **critical point**, where

$$\left.\frac{\partial^3 g}{\partial x_A^3}\right|_{T,P} = 0 , \tag{3.31}$$

because at the critical point the curvature obviously changes sign.

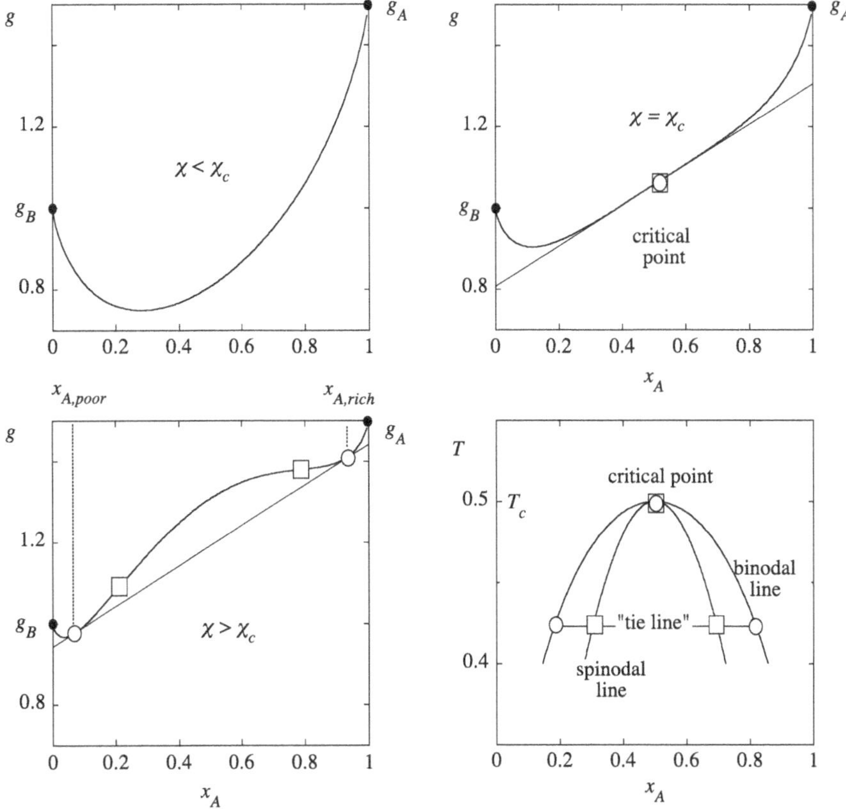

Fig. 3.3 Schematic of the x_A-dependence of g for different χ-values. The curves shown here are for $\chi = 1$ (upper left), $\chi = 2$ (upper right), and $\chi = 3$ (lower left) using $g_A = 1.5$ and $g_B = 1.0$. Lower right: T-x_A-phase diagram of our model of a binary mixture, where we assume that $T = 1/\chi$

Note that temperature here enters via the assumed proportionality $\chi \propto 1/T$. This assumption accounts for the observation that phase separation often occurs upon lowering temperature. Nevertheless this is purely empirical and more complex descriptions of χ can be found.

Figure 3.4 shows experimental liquid-liquid equilibria data for the binary mixtures water/phenol (solid squares) and methanol/hexane (solid circles) (data from the [8]). Here x_A is the mole fraction of water and methanol, respectively. Notice that while both systems show the basic behavior predicted by our theory, only the second system also exhibits the symmetry with respect to $x_A = 0.5$. Nevertheless, the solid lines are theoretical results, which where obtained using

$$\chi = \frac{c_0 + c_1 x_A}{T} + c_2 \; . \tag{3.32}$$

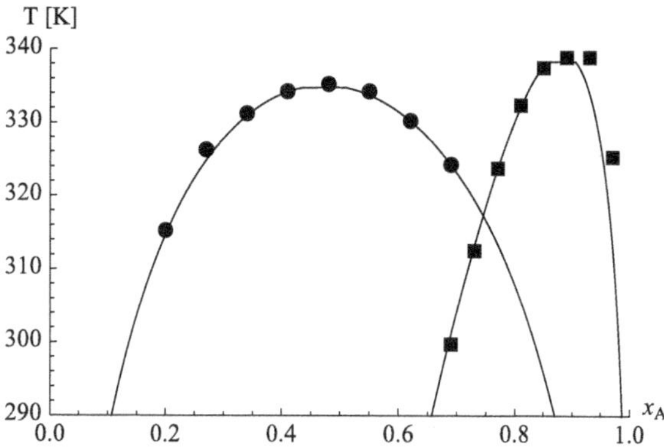

Fig. 3.4 Liquid-liquid equilibria data for the binary mixtures methanol/hexane (solid circles) and water/phenol (solid squares)

Here c_0, c_1, and c_2 are constants, which are adjusted so that the theory matches the data points. In particular the c_1-term breaks the symmetry around $x_A = 0.5$. While it is quite common to introduce such expressions for χ, it is not easy to provide reasonable physical explanations for the individual terms. In addition, the 'best fit' usually does not correspond to a unique set of values for c_0, c_1, and c_2.

However, let us return to polymer mixtures. Figure 3.5 shows analogous binodal data points for a macromolecular fluid mixture determined by observation of the **cloud points**. The term 'cloud point' refers to the turbidity observed upon passing from the homogeneous mixture into the coexistence region, where droplet formation increases the scattering of light. In our theoretical description we assume a fully occupied lattice.

The aforementioned figure shows polystyrene-polybutadiene mixture cloud point data taken from Fig. 3 in [9]; PS2-PBD2 (solid circles); PS3-PBD2 (open squares), and PS5-PBD26 (open triangles); $m_{PS2} = 2220$, $m_{PS3} = 3500$, $m_{PS5} = 5200$, $m_{PBD2} = 2350$, $m_{PBD26} = 25000$. Note that ϕ_1 refers to PS. The solid lines are the results of a calculation analogous to the one which produced the solid lines in Fig. 3.4, i.e. we determine the binodal line by the **common tangent construction** applied to the **mixing free enthalpy**

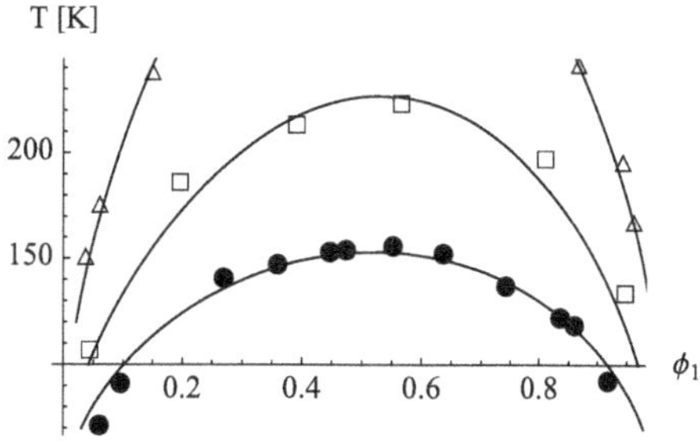

Fig. 3.5 Binodal data points and theoretical fits for three binary polymer mixtures

$$\frac{\Delta G}{N k_B T} = \frac{\phi_1}{m_1} \ln \phi_1 + \frac{1 - \phi_1}{m_2} \ln(1 - \phi_1) + \chi \phi_1 (1 - \phi_1) \tag{3.33}$$

(It does not matter whether we apply the common tangent construction to the mixing free enthalpy or to the full free enthalpy.). Again we use (3.32) to describe χ. Note that the χ-term in the literature sometime is denoted as an enthalpic contribution. This is not necessarily true, as we have already pointed out, because $\partial G/\partial T|_P = -S$, and if χ depends on temperature, as it usually does, then the χ-term contributes to the entropy as well. Here significant insight is needed into the microscopic interaction of polymer systems. A good starting point for the interested reader is [5].

3.3 Polymers in Solution

We briefly want to discuss (3.33) when $m_2 = 1$, i.e.

$$\frac{\Delta G}{n R T} = \frac{\phi}{m} \ln \phi + (1 - \phi) \ln(1 - \phi) + \chi \phi (1 - \phi) , \tag{3.34}$$

where $\phi_1 = \phi$. This situation describes a polymer-solvent-system. The phase behavior of this system is in principle described by Fig. 3.3 (bottom-right panel), except of course without the symmetry around $x_A = 0.5$, i.e. $\phi_1 = 0.5$, unless $m_1 = 1$. Setting the coefficient $c_1 = 0$ in (3.32) we easily work out the critical temperature and the critical packing fraction, i.e.

$$\frac{c_0}{T_c} = \frac{1}{2} - c_2 + \frac{1}{\sqrt{m}} + \frac{1}{2m} \quad \text{and} \quad \phi_c = \frac{1}{\sqrt{m} + 1} , \tag{3.35}$$

according to Shultz and Flory [10] (Note that T_c and ρ_c do agree with the same critical parameters in the case of the previously discussed gas-liquid critical point, cf. (3.20) and (3.21), if $c_o = N_A \epsilon_o/(2R)$ and $c_2 = 0$.). These critical parameters follow via the simultaneous solution of

$$\frac{\partial^2}{\partial \phi^2} \Delta G = 0 \quad \text{and} \quad \frac{\partial^3}{\partial \phi^3} \Delta G = 0 \tag{3.36}$$

(cf. (3.30) and (3.31)). The critical solution temperature, T_c, measured for different m may be fitted via (3.35), i.e. T_c^{-1} versus $m^{-1/2} + (2m)^{-1}$, to determine c_0 and c_2 experimentally (for this particular mixture).

3.4 Osmotic Pressure in Polymer Solutions

Equation (3.34) may be used to calculate the **osmotic pressure** Π of polymers in solution. However, here we want to use (3.33) instead, where m_2 is not yet equal to one. We employ the **Gibbs–Duhem equation** at constant temperature to obtain

$$V\Pi_1 = \int_0^{\nu_1} d\mu_1(\nu_1')\nu_1' \tag{3.37}$$

$$= \int_0^{\nu_1} d\nu_1'\nu_1' \frac{\partial^2 \Delta G}{\partial \nu_1^2}\bigg|_{\nu_1=\nu_1'} = -\frac{\partial \Delta G/\nu_1}{\partial(1/\nu_1)}$$

(see Problem 3.4). Note that $d\mu_1(\nu_1)$ is due to altering the relative polymer content of the solution, which solely affects the mixing contribution of the free enthalpy. After some work, using $n = N/N_A$, $N = m_1\nu_1 + m_2\nu_2$, and $\phi_i/m_i = \nu_i/(nN_A)$, we find

$$\Pi_1 = \frac{RT}{V}n_2\left(-\phi_1\left[1-\frac{m_2}{m_1}\right] - \ln(1-\phi_1) - m_2\chi\phi_1^2\right), \tag{3.38}$$

where $n_i = \nu_i/N_A$ $(i = 1, 2)$. It is instructive to expand the right hand side for small ϕ_1:

$$\Pi_1 = \frac{k_BT}{b^3}\left(\frac{\phi_1}{m_1} + \left[\frac{1}{2m_2}-\chi\right]\phi_1^2 + \frac{1}{3m_2}\phi_1^3 + \mathcal{O}(\phi_1^4)\right). \tag{3.39}$$

The quantity b^3 is the cell volume. Note that the derivation of (3.37) includes the assumption that the volume V does not change when the solute concentration increases. In the above expansion this must be assumed as well in order to be consistent with the derivation of (3.37) (see again Problem 3.4). This means that $V = b^3 m_2\nu_2$.

We do not want to discuss (3.38) and (3.39) in much detail. A few remarks must suffice:

Remark 1: Note that the first term in (3.39) is proportional to the number concentration polymer, whereas already the second term is proportional the square of the weight concentration. This is indeed in accord with Fig. 1.6. However, the obvious question is this: How well does (3.38) describe the data in Fig. 1.6 in general? We discuss the answer in Problem 3.5 based on a scaling theory of osmotic pressure.

Remark 2: From (3.39) we also conclude that the quantity $B_2 = A_2\bar{M}_w^2$ in (2.112) is given by

$$B_2 = b^3\left[\frac{1}{2}-\chi\right]m_1^2. \tag{3.40}$$

Remark 3: The term in square brackets in the (3.39) is proportional to the second virial coefficient of type-1-molecules in type-2-molecules. Suppose that the type-2-molecules are also chains. In fact they are the same type of polymer as type 1—maybe just somewhat shorter to make them a little different. In this case we expect that χ (cf. (3.10)) is very small. Since $1/(2m_2)$ is also small when the type-2-molecules are chains, we conclude that the entire second virial coefficient is small. Essentially we are looking at a situation which is quite similar to our discussion of the θ-solvent in the context of (2.74). This equation led us to conclude that the exponent ν, in a situation when the second viral coefficient vanishes, assumes its ideal chain-value. Following this line of reasoning we can see why polymer chains in a melt of identical chains should be ideal.

An insightful discussion of osmotic pressure in polymer solutions, including the present result, can be found in [11].

3.5 Swelling of Polymer Networks

Equation (3.34) describes the mixing free enthalpy of a polymer-solvent system on a lattice, where ϕ is the volume fraction polymer. If we express the solvent volume fraction $1 - \phi$ via $1 - \phi = N_s/N$, where N_s is the number of solvent cells and N is the total number of cells, the result is

$$\frac{\Delta G}{k_B T} = \nu \ln \phi + N_s \ln(1 - \phi) + \chi \phi N_s . \tag{3.41}$$

The quantity ν is the number of polymer chains (not to be confused with the exponent ν!). How can we adopt this equation to the swelling of a **polymer network**?

What we mean by 'network' is depicted in Fig. 3.6. Panel (a) shows a cartoon of several polymer chains or segments joined with other chains at their ends. These junction points are cross-links. Typically four chain segments are joined together in one cross-link. This happens for instance due to a covalent bridge between two linear chains. Aside from chemical cross-links two polymer chains may be entangled when one is making a hairpin-type bend around the other. In the following we are not interested in the true nature of the cross-link. We merely assume that the polymer network is a network of ν cross-linked chain segments containing n Kuhn lengths each. In addition, if a macroscopic volume element inside the network, possessing the edge lengths L_x, L_y, and L_z, is deformed during swelling, then its new edge lengths are $L'_x = \lambda_x L_x$, $L'_y = \lambda_y L_y$, and $L'_z = \lambda_z L_z$. We also assume that the segment end-to-end vectors contained in this volume element will change their components analogously. Hence

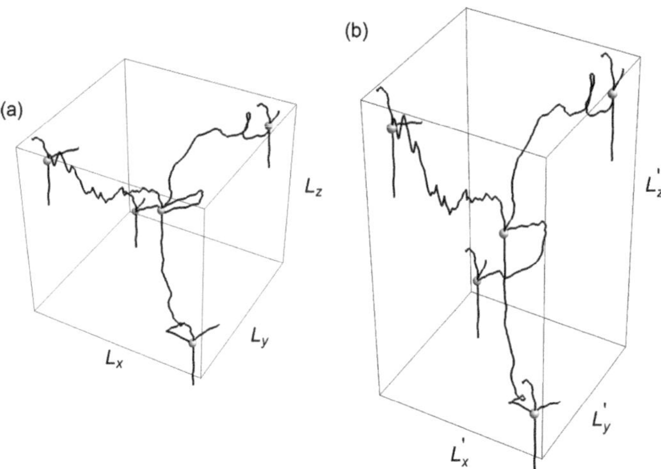

Fig. 3.6 A polymer network volume element containing cross-linked chain segments in the undeformed (**a**) and in the deformed (**b**) state. Cross-links are indicated by the dots

$$\frac{L_\alpha}{R_{i,\alpha}} = \frac{L_\alpha'}{R_{i,\alpha}'} , \qquad (3.42)$$

which is an affine deformation (cf. Fig. 3.6b).

Here we want to study uniform swelling, i.e. $\lambda_x = \lambda_y = \lambda_z \equiv \lambda > 1$. But there is another important case, which we shall discuss as well. **Elastomers** are a certain class of polymers commonly used to manufacture **rubber**. Rubber has a very small compressibility, i.e. $V' = L_x' L_y' L_z' \approx L_x L_y L_z = V$. When we stretch a rubber band by a factor $\lambda_z \equiv \lambda$ in z-direction, i.e. $L_z' = \lambda L_z$, **volume conservation**, i.e. $\lambda_x \lambda_y \lambda_z = 1$, implies $\lambda_x = \lambda_y = \lambda^{-1/2}$.

Let us come back to our original question—how must we modify (3.41) in order to describe the free enthalpy change during swelling of a polymer network in contact with a solvent? In a simple approximate theory of polymer network swelling due to Flory and Rehner [12] the **translational entropy** term $\nu \ln \phi$ is replaced by an **elastic entropy** contribution $\frac{3\nu}{2}(\lambda^2 - 1 - \ln \lambda)$ when the network undergoes uniform swelling, i.e. $\lambda_x = \lambda_y = \lambda_z \equiv \lambda$. Note that the cross-links provide a framework which fixes the positions of the chain segments in space, i.e. the translational entropy does not change for these segments. However, their end-to-end distance changes during swelling (or, more generally, during the deformation) of the network. Thus, we may remove the term which arises from the number of ways in which we may place the polymer chains on the lattice. It is replaced by the entropy change (times $-T$) due to changes of the chain's end-to-end distances.

Why this entropy change has the above form, we shall discuss in detail below. Momentarily we just assume that this is indeed the correct form and proceed. Minimizing the new ΔG with respect to N_s, i.e. $0 = \partial(\Delta G/(k_B T))/\partial N_s$, which means

equating the solvent chemical potentials inside and outside the network, yields the
Flory–Rehner equation in its standard form:

$$\rho_P \frac{\upsilon_s}{m_P} = -\frac{\ln(1-\phi) + \phi + \chi\phi^2}{\phi^{1/3} - \phi/2} . \tag{3.43}$$

The quantity ρ_P is the mass density of the (dry) polymer, υ_s is the molecular volume
of the solvent, and m_P is the (average) mass of a network segment. Note that $\lambda^3 =
V_{swollen}/V_P = 1/\phi = (V_P + N_s\upsilon_s)/V_P$, where $V_{swollen}$ is the equilibrium volume
of the swollen network (or gel) and V_P is the dry polymer volume. Note also that
$\partial\phi/\partial N_s = -\phi^2\upsilon_s/V_p$ (caveat: It is incorrect to use the above formula $\phi = 1 -
N_s/N$, assuming $N = const$, for the derivative of ϕ with respect to N_s. This means
that the volume is kept constant—which is not the case.), $\partial\lambda/\partial N_s = \upsilon_s/(3V_P\lambda^2)$, as
well as $\rho_P\frac{\upsilon_s}{m_P} = \frac{\upsilon_s}{V_P}\nu$.

In the original paper, cf. (11) in [12], the $-\phi/2$-term in the denominator of (3.43),
which results from the $-\ln\lambda$-term in the elastic entropy, is not included. This per-
sisted for some time (e.g., (5.9) in [13]), but finally 'converged' to the above form.
However, the significance of the $-\phi/2$-term is somewhat questionable. The factor
$1/2$ resulted from a rather approximate entropy contribution of the network nodes
and, in addition, is based on the assumption of a regular 4-fold coordinated network.
In many cases of practical importance $\phi^{1/3}$ dominates over $\phi/2$ and the latter can be
neglected. Nevertheless, (3.43) ties the quantity ν to the swollen volume of a polymer
network. Therefore it provides another means for the determination of the cross-link
density (see also the footnote in the context of (4.20)).

While the Flory–Rehner equation is used as a standard means to deduce m_P and
hence the cross-link density by measuring ϕ, this is not really a test of (3.43). It is
therefore interesting to verify the Flory–Rehner equation using computer simulations, which allow to determine the cross-link density as well as ϕ independently
(see for example [14]).

- Derivation of the elastic entropy contribution $\frac{3\nu}{2}(\lambda^2 - 1 - \ln\lambda)$:

The entropy change ΔS^{el} due to the deformation of the network has two contributions,
i.e.

$$\Delta S^{el} = \Delta S^{el,f} + \Delta S^{el,\nu} . \tag{3.44}$$

$\Delta S^{el,f}$ is the entropy change relative to the undeformed state without considering
cross-links explicitly. $\Delta S^{el,\nu}$ is a correction due to the explicit presence of cross-links.
We start with $\Delta S^{el,f}$, which is given by

$$\Delta S^{el,f} = k_B \ln\left[\frac{\Omega'}{\Omega}\right] . \tag{3.45}$$

The quantity Ω is the number of different ways in which N polymer segments can be accommodated inside the undeformed volume V. Here and in the following primes indicate corresponding quantities in the deformed state. Note that we group the segments according to their end-to-end vectors $\vec{R}_i$. Hence

$$\Omega \propto \frac{N!}{\prod_i N_i!} \prod_i p_i^{N_i} \,. \tag{3.46}$$

Here $p_i^{N_i}$ is the probability of finding N_i polymer segments with their chain ends in the same volume element at $\vec{R}_i$ (all segments start at a common origin)—with p_i being this probability for a single segment. The factor $N!(\prod_i N_i!)^{-1}$ accounts for the distinguishable permutations of segments provided that their distinguishing feature is their end-to-end vector (i.e. the volume element it points to). The deformation redistributes the segments, i.e. $N_i \to N_i'$. Hence

$$\Omega' \propto \frac{N!}{\prod_i N_i'!} \prod_i p_i^{N_i'} \,.$$

Note that p_i does not change! p_i is the probability density $p(R_i)$ in (2.67) multiplied by the volume element at $\vec{R}_i$. Therefore p_i is the ratio of the number of conformations which let a single polymer segment end up in the volume element at $\vec{R}_i$ divided by the number of all its possible conformations. This ratio, which is based on the intrinsic polymer properties, is not altered by volume deformations provided that we keep looking at the same volume element regardless of the deformation state. What changes though is the number of segment endpoints N_i in this volume element. Swelling for instance will dilute these points and reduce their number to $N_i' < N_i$.

Now we apply Stirling's approximation to $N_i!$ and $N_i'!$, i.e. $\ln N_i! \approx N_i \ln N_i - N_i$ and $\ln N_i'! \approx N_i' \ln N_i' - N_i'$. In addition we use $p_i = N_i/N$, $p_i' = N_i'/N$ and $\sum_i N_i = \sum_i N_i' = N$. The result is

$$\ln\left[\frac{\Omega'}{\Omega}\right] = N \sum_i p_i' \ln\left[\frac{p_i}{p_i'}\right] \,. \tag{3.47}$$

Let's briefly discuss $p_i = N_i/N$ and $p_i' = N_i'/N$. The first equation applies in the undeformed state and is (hopefully) evident. As a definition of p_i' the second equation is evident also. But what maybe confusing is the difference between p_i' and p_i. As we had already mentioned, p_i is the probability density $p(R_i)$ in (2.67) multiplied by the volume element at $\vec{R}_i$, i.e.

$$p_i = \Delta R_{i,x} \Delta R_{i,y} \Delta R_{i,z} \frac{c^3}{\pi^{3/2}} \exp\left[-c^2 \left(R_{i,x}^2 + R_{i,y}^2 + R_{i,z}^2\right)\right] \,, \tag{3.48}$$

where $c^2 \equiv (2\langle \vec{R}^2 \rangle/3)^{-1}$. Network deformation means that the old and the new coordinates are related via

$$R_{i,\alpha} = \frac{1}{\lambda_\alpha} R'_{i,\alpha} \quad (\alpha = x, y, z) . \tag{3.49}$$

Inserting this into (3.48) yields

$$p'_i = \Delta R'_{i,x} \Delta R'_{i,y} \Delta R'_{i,z} \frac{c^3}{\pi^{3/2} \lambda_x \lambda_y \lambda_z} \tag{3.50}$$
$$\times \exp\left[-c^2 \left((R'_{i,x}/\lambda_x)^2 + (R'_{i,y}/\lambda_y)^2 + (R'_{i,z}/\lambda_z)^2\right)\right]$$

and (3.47) becomes

$$\ln\left[\frac{\Omega'}{\Omega}\right] = N \ln\left[\lambda_x \lambda_y \lambda_z\right] - N \sum_i c^2 p'_i \left[R'_{i,x}{}^2 \left(1 - \frac{1}{\lambda_x^2}\right) + \dots\right]$$
$$= N \ln\left[\lambda_x \lambda_y \lambda_z\right]$$
$$- N \int dR'_x dR'_y dR'_z \frac{c^3 \exp\left[-c^2 \left((R'_x/\lambda_x)^2 + \dots\right)\right]}{\pi^{3/2} \lambda_x \lambda_y \lambda_z}$$
$$\times \left[c^2 R'_x{}^2 \left(1 - \frac{1}{\lambda_x^2}\right) + \dots\right]$$
$$= N \ln\left[\lambda_x \lambda_y \lambda_z\right] - \frac{N}{2} \left(\lambda_x^2 + \lambda_y^2 + \lambda_z^2 - 3\right)$$

or

$$\Delta S^{el,f} = N k_B \ln\left[\lambda_x \lambda_y \lambda_z\right] - \frac{N k_B}{2} \left(\lambda_x^2 + \lambda_y^2 + \lambda_z^2 - 3\right) . \tag{3.51}$$

Hence, for a uniform volume change $V' = \lambda^3 V$ we find

$$\Delta S_{iso}^{el,f} = 3 N k_B \ln \lambda - \frac{3 N k_B}{2} \left(\lambda^2 - 1\right) , \tag{3.52}$$

whereas for a volume conserving uniaxial stretch ($\lambda_z = \lambda$, $\lambda_x = \lambda_y = \lambda^{-1/2}$)

$$\Delta S_{stretch}^{el,f} = -\frac{N k_B}{2} \left(\lambda^2 + \frac{2}{\lambda} - 3\right) . \tag{3.53}$$

Figure 3.7 shows $\Delta S_{iso}^{el,f}/(N k_B)$ and $\Delta S_{stretch}^{el,f}/(N k_B)$ versus λ. Any deviation from the undeformed state $\lambda = 1$ results in a negative entropy contribution. When $\lambda > 1$ the chain segments are stretched in both situations and the number of possible chain conformations are reduced. When $\lambda < 1$ it is the 'ideal gas volume term' $\ln \lambda^3$, which causes the entropy change to become negative in the case of $\Delta S_{iso}^{el,f}/(N k_B)$.

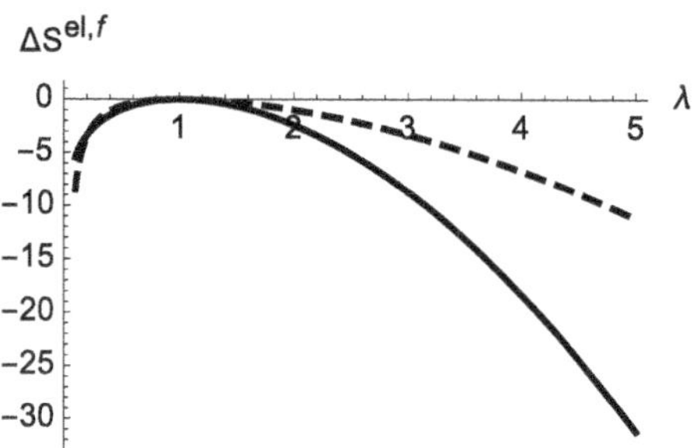

Fig. 3.7 $\Delta S^{el,f}_{iso}/(Nk_B)$ (solid line) and $\Delta S^{el,f}_{stretch}/(Nk_B)$ (dashed line) versus λ

In the volume conserving case of $\Delta S^{el,f}_{stretch}/(Nk_B)$ a reduction of λ below one again means chain stretching albeit in the xy-plane.

Next we focus on $\Delta S^{el,v}$, the correction to ΔS^{el} due to the cross-links. Let's assume a network consisting of N segments. In the most common situation we have four segments 'emanating' from one cross-link. Since only half of each segment belongs to this cross-link (the other half belongs to the next cross-link), there are two full segments per cross-link or $N/2$ cross-links total.

Flory has argued that there is an entropy change, i.e. $\Delta S^{el,v}$, originating from the confinement of each cross-link to essentially a small volume δV due to the network structure. We estimate this entropy change via

$$\Delta S^{el,v} = k_B \ln \frac{(\delta V/V')^{N/2}}{(\delta V/V)^{N/2}} = -\frac{Nk_B}{2} \ln\left[\lambda_x \lambda_y \lambda_z\right] . \tag{3.54}$$

Note that $\delta V/V$ and $\delta V/V'$ is the probability for finding a cross-link in the small volume δV in the undeformed and in the deformed state, respectively. Assuming that the $N/2$ cross-links are independent, which is rather crude, we obtain the first equation in (3.54). Note also that this contribution to the entropy does not enter if the volume is conserved, i.e. $\Delta S^{el,f}_{stretch}$ is the total $\Delta S^{el}_{stretch}$. In other words,

$$\Delta S^{el}_{iso} = 3Nk_B\left(\ln\lambda - \lambda^2 + 1\right) \tag{3.55}$$

and

$$\Delta S^{el}_{stretch} = -\frac{Nk_B}{2}\left(\lambda^2 + \frac{2}{\lambda} - 3\right) . \tag{3.56}$$

In general the total ΔS^{el} becomes

$$\Delta S^{el} = \frac{Nk_B}{2}\left(\ln\left[\lambda_x\lambda_y\lambda_z\right] - \lambda_x^2 - \lambda_y^2 - \lambda_z^2 + 3\right) . \tag{3.57}$$

With an extra factor $-T$, $\lambda = \lambda_x = \lambda_y = \lambda_z$, and N replaced by v this is exactly the term which replaces $Tv\ln\phi$ in (3.41).

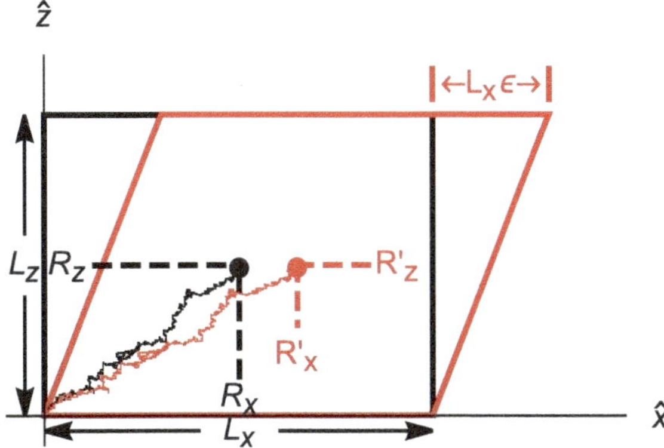

Fig. 3.8 Uniform shearing of a block of material

Before moving on we want to address the important case of ΔS^{el} due to uniform shear depicted in Fig. 3.8. Again $V = const$. Instead of (3.49) we now have

$$R_{i,x} = R'_{i,x} - \gamma_{xz} R'_{i,z} \quad (\gamma_{xz} \equiv \frac{L_x}{L_z}\epsilon)$$

$$R_{i,y} = R'_{i,y} \tag{3.58}$$

$$R_{i,z} = R'_{i,z} \, ,$$

which, after insertion into the expression for $\ln[\Omega'/\Omega]$, yields

$$\Delta S^{el} = \Delta S^{el,f}_{shear} = -\frac{Nk_B}{2}\gamma_{xz}^2 \, . \tag{3.59}$$

3.6 Problems

Problem 3.1 Derive Stirling's formula, i.e. (3.3), via the following steps:

(a) Show first that

$$N! = \Gamma[N+1] \, , \tag{3.60}$$

where

$$\Gamma[x] = \int_0^\infty e^{-t} t^{x-1} dt \tag{3.61}$$

is the **Gamma function**. Hint: Show $\Gamma[x+1] = x\,\Gamma[x]$ and $\Gamma[1] = 1$.

(b) Express $\Gamma[x + 1]$ via

$$\Gamma[x + 1] = \int_0^\infty e^{Nf(t)} dt \tag{3.62}$$

and plot $f(t)$, for example, using $N = 10$. You will find that $f(t)$ has a maximum. Note that the vicinity of this maximum yields the largest contribution to $\Gamma[x + 1]$. Now work out the first two terms of a Taylor expansion of $f(t)$ at its maximum.

(c) Calculate $\Gamma[x + 1]$ using the expansion of $f(t)$ obtained in (b). Taking the logarithm on both sides of the resulting equation yields (3.3).

Problem 3.2 (a) Show that (3.24) is correct.

(b) Show that the critical parameters T_c and ϕ_c in the (3.20) and (3.21) can be obtained from the first three terms in the expansion of the equation of state (3.19) in small ϕ when m is large.

Problem 3.3   Recreate the plots in the figures (3.3), (3.4), and (3.33).

Problem 3.4 Derive (3.37) for the osmotic pressure in polymer solutions. Begin with a short derivation of the Gibbs–Duhem equation and continue by filling in the missing steps.

Problem 3.5 We want to construct a scaling theory for the osmotic pressure in polymer solutions beyond the (monomer) overlap concentration ρ^* (cf. the end of Chap. 2) and apply it to the data in Fig. 1.6.
   Our starting point is van't Hoff's law (1.6), i.e.

$$\frac{N\Pi}{\rho k_B T} = 1 . \tag{3.63}$$

In order to arrive at this form of (1.6) we use $c = m\rho$, where m is the monomer mass, and $\bar{M}_n = Nm$, where N is the number of monomers per (monodisperse) polymer. Now we want to extend this formula, which is valid in the limit of small ρ, to 'high' monomer concentrations. The only density apparently particular to this problem, separating the two regimes, is the **overlap concentration** ρ^*. Hence we express the monomer density ρ in units of ρ^* and study the regime $\rho/\rho^* > 1$.

(a) Provide a rough estimate of $c^* = m\rho^*$ for the three upper data sets (denoted α-104, α-12, and α-103 from top to bottom in the original paper) in Fig. 1.6 (the same data sets we had used in Problem 1.2). Do this for $\nu = 1/2$ as well as for $\nu = 3/5$.

(b) Try the ansatz

$$\frac{N\Pi}{\rho k_B T} = Z(\rho/\rho^*) , \qquad (3.64)$$

where $Z(\rho/\rho^*)$ is an unknown **scaling function**. Since the data in the figure suggest that $Z(\rho/\rho^*)$ may be a power law, we also try $Z(\rho/\rho^*) \sim (\rho/\rho^*)^x$. Find x using the condition that the osmotic pressure at high concentrations does not depend on N.

(c) Get the original three data sets from Table II in [15] and show that $x = 1.25$ yields an excellent fit to the data at large c/c^* by plotting $\bar{M}_n \Pi/(cRT)$ versus c/c^*.

References

1. M. Abramowitz, I.A. Stegun (eds.), *Handbook of Mathematical Functions* (Dover, New York, 1972)
2. A.J. Stavermann, J.H. van Santen, Rec. Trav. Chim. **60**, 76 (1941)
3. M.L. Huggins, Solutions of long chain compounds. J. Chem. Phys. **9**, 440 (1941); Thermodynamic properties of solutions of long-chain compounds. Ann. NY Acad. Sci. **43**, 1 (1942)
4. P.J. Flory, Thermodynamics of high polymer solutions. J. Chem. Phys. **9**, 660 (1941); **10**, 51 (1942)
5. R. Koningsveld, L.A. Kleintjens, Fluid phase equilibria in macromolecular systems. Acta Polymerica **39**, 341 (1988)
6. E.D. Nikitin, The critical properties of thermally unstable substances: measurement methods, some results and correlations. High Temp. **36**, 305 (1998)
7. R. Hentschke, *Thermodynamics*, 2nd edn. (Springer, Heidelberg, 2022)
8. D.R. Lide, H.V. Kehiaian (eds.), *CRC Handbook of Thermophysical and Thermochemical Data* (CRC Press, Boca Raton, 1994)
9. R.-J. Roe, W.-C. Zin, Determination of the polymer-polymer interaction parameter for the polystyrene-polybutadiene pair. Macromolecules **13**, 1221 (1980)
10. A.R. Shultz, P.J. Flory, Phase equilibria in polymer-solvent systems. J. Am. Chem. Soc. **74**, 4760 (1952)
11. P.-G. de Gennes, *Scaling Concepts in Polymer Physics* (Cornell University Press, Ithaca, 1988)
12. P.J. Flory, J. Rehner, *Statistical mechanics of cross linked polymer networks II. Swelling.* J. Chem. Phys. **11**, 521 (1943)
13. L.R.G. Treloar, The elasticity and related properties of rubbers. Rep. Progress Phys. **36**, 755 (1973)
14. M. Lang, A. John, J.-U. Sommer, Model simulations on network formation and swelling as obtained from cross-linking co-polymerization reactions. Polymer **82**, 138 (2016)
15. I. Noda, N. Kato, T. Kitano, M. Nagasawa, Thermodynamic properties of moderately concentrated solutions of linear polymers. Macromolecules **16**, 668 (1981)

Chapter 4
Polymer Dynamics

Abstract In the following we introduce time as a variable and study its relation to other quantities like a polymer's mass or its temperature. We begin by developing theoretical descriptions for viscoelasticity, both phenomenological as well as microscopic. We then move on to a discussion of the glass process. Here our focus is on empirical findings, but we shall also look into one of the theories originally developed in this context.

4.1 Linear Deformation Mechanics

This subsection applies mainly to polymer networks and gels. Experiments studying the time-dependent behavior of these systems often involve the mechanical deformation of bulk samples. Even though this is a somewhat narrow scope, there are a number of concepts and quantities with broader importance addressed here, e.g., storage and loss modulus, complex modulus, compliance, etc. It is this which justifies the somewhat cursory discussion of linear deformation mechanics at this point. On a first reading the 'boxes' *Equations and Concepts from Isotropic Elasticity* (cf. L. D. Landau, E. M. Lifshitz *Theory of Elasticity*) and *Equations and Concepts from Fluid Mechanics* (cf. L. D. Landau, E. M. Lifshitz *Fluid Mechanics*) may be skipped over. However, both boxes contain essential physics underlying what is called **'viscoelasticity'**, the main feature of polymers in motion, which means that they should not be neglected entirely.

Equations and Concepts from Isotropic Elasticity:

Deformation of an elastic body will displace a point at $\vec{r}$ in its interior by $\vec{u}(\vec{r})$. This means that the new position $\vec{r}\,'$ is

$$\vec{r}\,' = \vec{r} + \vec{u}(\vec{r}). \tag{4.1}$$

If two points inside an undeformed sample are infinitesimally close, the vector from one to the other, $d\vec{r}$, transforms according to

$$d\vec{r}\,' = d\vec{r} + d\vec{u}(\vec{r}) = d\vec{r} + d\vec{r} \cdot \vec{\nabla}\vec{u}(\vec{r}). \tag{4.2}$$

Hence, the square of $d\vec{r}\,'$ is

$$dr'^2 = (dx_i + dx_j \partial_j u_i)^2 \approx dx_i^2 + 2\,dx_i dx_j \partial_j u_i \tag{4.3}$$

($\partial_j \equiv \partial/\partial x_j$). Here we use the summation convention and we assume that $\vec{u}(\vec{r})$) varies slowly in space. The last term in (4.3) can be written as

$$2\,dx_i dx_j \partial_j u_i = 2\,dx_i dx_j \frac{1}{2}(\partial_j u_i + \partial_i u_j) \equiv 2\,dx_i dx_j\, u_{ij}. \tag{4.4}$$

This defines the components of the **strain tensor**, i.e.

$$u_{ij} = \frac{1}{2}\left(\frac{\partial u_i}{\partial x_j} + \frac{\partial u_j}{\partial x_i}\right). \tag{4.5}$$

Note that the strain tensor is the central quantity here. It contains all the local information regarding the deformation.

Since the strain tensor is symmetric, i.e. $u_{ij} = u_{ji}$, it may be diagonalized at every point:

$$
\begin{aligned}
dr'^2 &= \left(\delta_{ij} + 2u_{ij}\right)dx_i dx_j \\
&= \left(1 + 2u^{(1)}\right)dx^2 + \left(1 + 2u^{(2)}\right)dy^2 + \left(1 + 2u^{(3)}\right)dz^2.
\end{aligned}
$$

The quantities $u^{(i)}$ are the attendant principal values or eigenvalues of the strain tensor. Therefore

$$dx_i' = \sqrt{\left(1 + 2u^{(i)}\right)}dx_i. \tag{4.6}$$

If $dV' = dx'dy'dz'$ is the volume element in the deformed state and $dV = dxdydz$ is the volume element in the undeformed state, we can use (4.6) to obtain

$$
\begin{aligned}
dV' &= \sqrt{1 + 2u^{(1)}}\sqrt{1 + 2u^{(2)}}\sqrt{1 + 2u^{(3)}}dx_1 dx_2 dx_3 \\
&\approx \left(1 + u^{(1)}\right)\left(1 + u^{(2)}\right)\left(1 + u^{(3)}\right)dV \\
&\approx \left(1 + u^{(1)} + u^{(2)} + u^{(3)}\right)dV.
\end{aligned}
\tag{4.7}
$$

The quantity $u^{(1)} + u^{(2)} + u^{(3)}$ is the trace of the strain tensor. Because the trace is independent of the coordinate system, we find

$$dV' = (1 + u_{ll})\, dV. \tag{4.8}$$

Note that a vanishing trace of the strain tensor implies local **volume conservation**.

Every deformation in the type of system we are interested in, i.e. an isotropic system, can be thought of as the being composed of a pure shear plus a pure dilation. Therefore it is useful to rewrite u_{ij} as

$$u_{ij} = \frac{1}{3}\delta_{ij}u_{ll} + \left(u_{ij} - \frac{1}{3}\delta_{ij}u_{ll}\right). \tag{4.9}$$

The trace of the term in brackets on the right hand side of this equation is zero ($\delta_{ll} = 3$). This term therefore describes a volume conserving shear deformation as depicted in Fig. 3.8. The other term describes a pure dilatation.

It is reasonable to assume that we can link u_{ij}, or more specifically these two terms individually, to a corresponding **elastic stress tensor** σ_{ij} via the linear equation

$$\sigma_{ij} = K\,\delta_{ij}u_{ll} + 2G\left(u_{ij} - \frac{1}{3}\delta_{ij}u_{ll}\right), \tag{4.10}$$

where K and G are constants. In order to find out whether our guess is sensible we study three special cases.

First we apply formula (4.10) to a gas and calculate the trace on both sides. We find

$$3\sigma = 3K u_{ll} \overset{(4.8)}{=} 3K\frac{\delta V}{V}. \tag{4.11}$$

Due to the nature of the system $\sigma \equiv \sigma_{xx} = \sigma_{yy} = \sigma_{zz}$. Comparing this to the formula for the isothermal compressibility, $\kappa_T = -V^{-1}\delta V/\delta P|_T$, we conclude $\sigma = -\delta P$ and $K = 1/\kappa$. Therefore σ_{ij} has units of force per area. In addition, K is the **compression modulus**.

Next we apply (4.10) to Fig. 3.8, which yields

$$\sigma_{xz} = 2G\,u_{xz} = G\frac{\partial u_x}{\partial z} = G\frac{L_x \epsilon}{L_z} \tag{4.12}$$

or

$$\sigma_{xz} = G\,\gamma_{xz}. \tag{4.13}$$

Hence, σ_{xz} is a force per area in x-direction applied to the face of the sample defined by its normal in z-direction and G is the **shear modulus**. If we assume that G is entirely due to the entropy elasticity of the chains in the sample, we can use our previous result (3.59) to obtain G via

$$\sigma_{xz} = \frac{1}{V}\frac{\partial(-T\Delta S^{(el)})}{\partial\gamma_{xz}} = \frac{N}{V}k_B T \gamma_{xz} \tag{4.14}$$

(Note that σ_{xz} is a local quantity referring to a certain volume element. Here all volume elements are identical and we divide simply by the total sample volume V.) or

$$G = \frac{N}{V}k_B T, \tag{4.15}$$

i.e. G is the chain density in the sample times $k_B T$.

The third case is uniaxial stretching or compression of the sample. Here it is useful to invert (4.10). We can do this by computing the trace, which yields $\sigma_{ll} = 3K u_{ll}$. Inserting $u_{ll} = \sigma_{ll}/(3K)$ back into (4.10) gives

$$u_{ij} = \frac{1}{9K}\delta_{ij}\sigma_{ll} + \frac{1}{2G}\left(\sigma_{ij} - \frac{1}{3}\delta_{ij}\sigma_{ll}\right). \tag{4.16}$$

We apply this equation to a stretched column depicted in Fig. 4.1. The forces stretching the column are distributed uniformly across the top and bottom faces of the column. No forces are applied to the other faces. This implies that the only component of the stress tensor which is not zero is σ_{zz}. Equation (4.16) tells us

Fig. 4.1 Stretching of a rectangular sample in z-direction

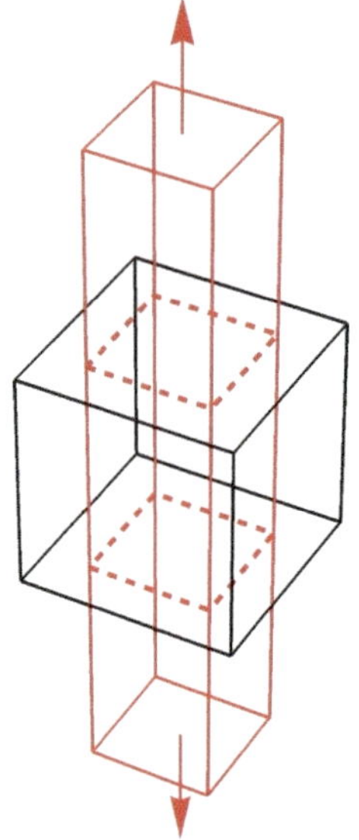

Table 4.1 Elastic modulus, shear modulus, and Poisson's ratio of selected materials

Material	E [GPa]	G [GPa]	ν
Iron	210	82	0.28
Aluminum	71	26	0.34
Polystyrene	3.2	1.2	0.35
Rubber, unfilled	$\approx 1 \cdot 10^{-3}$ to $5 \cdot 10^{-3}$	E/3	≈ 0.5
Rubber, filled	$\approx 1 \cdot 10^{-2}$ to $5 \cdot 10^{-2}$	E/3	≈ 0.5

therefore that $u_{ij} = 0$ if $i \neq j$. For the diagonal elements of the strain tensor we obtain

$$\sigma_{zz} = E\, u_{zz} \quad \text{and} \quad u_{xx} = u_{yy} = -\nu u_{zz}, \tag{4.17}$$

where

$$E = \frac{9KG}{3K + G} \quad \text{and} \quad \nu = \frac{1}{2}\frac{3K - 2G}{3K + G}. \tag{4.18}$$

The quantity E is the **elastic modulus** or **Young's modulus**, whereas ν is **Poisson's ratio** (again—do not confuse Poisson's ratio with the exponent ν we had introduced in (2.8).) (Table 4.1). In the limit when the volume does not change during a deformation, which means the compression modulus is very large—essentially infinite, this becomes

$$E = 3G \quad \text{and} \quad \nu = \frac{1}{2}. \tag{4.19}$$

We can compare the above to our previous result (3.56). The equation describes the entropy change in a sample containing N polymer chains when the sample is uniaxially stretched without volume change. Analogous to (4.14) we write

$$\sigma_{zz} = \frac{1}{V} \frac{\partial(-T \Delta S^{el}_{stretch})}{\partial \lambda} \tag{4.20}$$

$$= \frac{N}{V} k_B T \left(\lambda - \frac{1}{\lambda^2}\right) \overset{(\lambda \approx 1)}{\approx} 3\frac{N}{V} k_B T(\lambda - 1)$$

or

$$E = 3\frac{N}{V} k_B T \overset{(4.15)}{=} 3G. \tag{4.21}$$

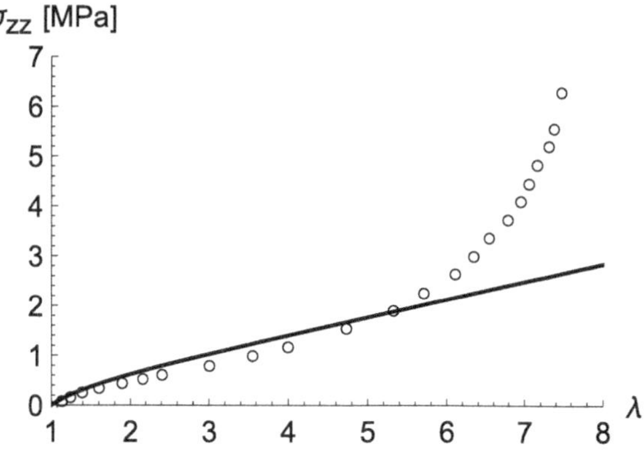

Fig. 4.2 Stress-strain curve of a highly vulcanized (8% S) natural rubber (NR) at 20 °C. The data are from Fig. 3 (graph (a)) in [2]. The solid line is $\sigma_{zz} = G(\lambda - \lambda^{-2})$ according to (4.20) fitted to the first three data points with $G \approx 0.36$ MPa

This[1] is a reassuring result in the sense that there appears to be consistency between the formalism for continuous isotropic elastic media in this subsection and our previous picture of elasticity based on conformation entropy of polymer chains.

Remark: There are different types of stress. We shall discuss them in Sect. 5.1 as well as in a problem. The reader should be aware of this, since it becomes important when formulas like (4.20) are fitted to experimental data.

Nevertheless, we should not be overly confident. The applicability of, for instance, (4.20) is limited as demonstrated by an example in Fig. 4.2. The description of entropy elasticity underlying (4.20) does not include the finite extensibility of real polymer chains. The latter gives rise to the marked increase of the stress. Even though this is the most obvious deviation, the rather insufficient inclusion of the network structure into the derivation that lead to (4.20) limits this equation to strains λ less than 1.5–2.

Finally, it is worth highlighting the following difference between (4.14) and (4.20). σ_{xz} depends linearly on γ_{xz}, independent of the magnitude of the shear strain. The same is not true for the λ-dependence of σ_{zz} during stretching of the sample, which is non-linear except in the limit $\lambda - 1 \ll 1$.

Remark 1: Figure 4.3 shows a measurement of stress versus temperature at constant strain for different strains using an elastomer sample. The results obtained for the larger strains meet our expectation, since the straight lines fitting the data exhibit positive slopes in accord with (4.20). At small strain, i.e. the lowest three of the

---

[1] Note that (4.20) as well as the **Flory–Rehner equation** (3.43), describing the equilibrium swelling of a polymer network, in principle can be used to measure the density of cross-links in the network (for a comparative overview of these and other methods see [1]). Both equations contain the number of polymer chains segments, defined as the (average) chain length between two successive cross-links as one moves along the polymer from one cross-link to the next. In praxis, however, (4.20), whose accuracy is quite limited, is replaced by a related one—the **Mooney–Rivlin equation** discussed in the section on aspects of the mechanics of polymers.

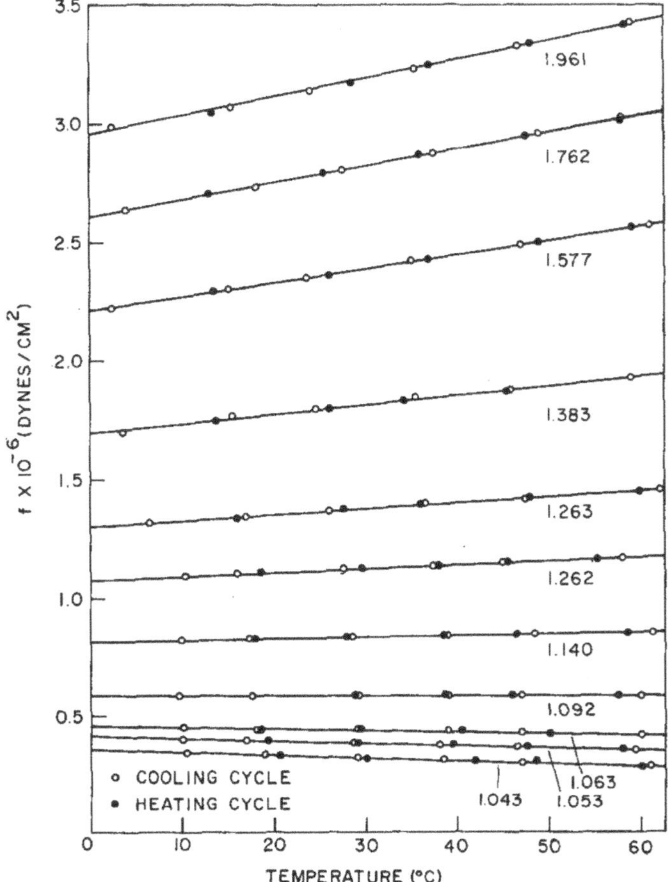

Fig. 4.3 Thermoelastic data of natural rubber. Numbers on the right-hand side of the graph indicate λ-values. This figure is reprinted with permission from [3]; *Note* 1 dyne = 10^{-5} N

curves, the slope is noticeably negative, in apparent disagreement with (4.20). How do we explain this?

The elastomer sample possess a positive coefficient of **thermal expansion**. Thermal expansion, however, is due to a piece in the elastomer's free energy, which is not part of the conformation entropy that is underlying (4.20). In the experiment both effects, the thermal expansion (causing expansion of the sample when it is heated) and entropy elasticity (causing contraction of the sample when it is heated), are present. At small strains the former dominates, whereas at large strains the latter dominates. The cross-over from one dominating to the other dominating is called **thermoelastic inversion**.

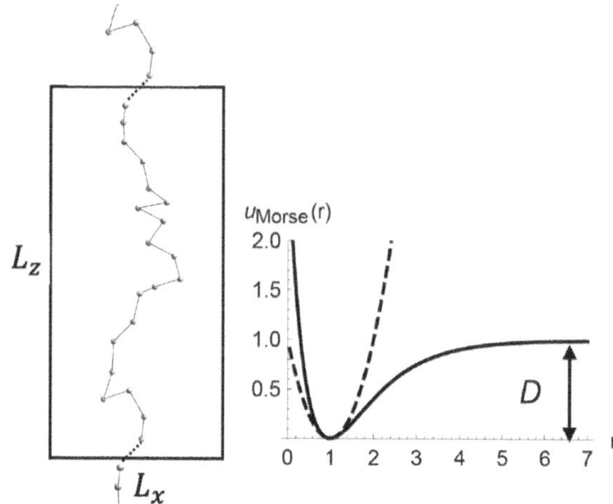

Fig. 4.4 Beads linked by Morse potentials (right) form a chain (left) which is periodically extended in z-direction. The dashed line is a harmonic approximation at the minimum of the Morse potential function

Remark 2—finite chain extension: Figures 4.4 and 4.5 illustrate finite chain extensibility by stretching a bead-spring model chain containing N beads in a **Langevin Dynamics** computer simulation. The latter simply means that we solve the equations of motion of the beads numerically using the Langevin equation which we shall discuss in Sect. 4.2. Neighboring beads along the chain are linked via **Morse potential** functions $u_{morse}(r) = D(1 - \exp[-a(r - r_o)])^2$ (here: $D = a = r_o = 1$). Note that the Morse function has a finite depth and therefore the Morse bonds can break. There is no other interaction between the beads. When the simulation box is stretched in z-direction so is the chain, which eventually ruptures. Initially the stress is governed by entropy elasticity. But as the elongation of the chain increases this contribution is reduced in comparison to the effect due to the strain on the individual bonds. Note that $L_{z,o}\lambda/N = 1$ corresponds to a value of λ at which the chain is fully stretched if the bond lengths are at their equilibrium value r_o. Details of this calculation can be found in [4]. It is also important to mention that the force required to stretch an isolated polymer chain can be measured in real experiments (see [5]).

It is prudent to note that this is not the standard approach to finite chain extensibility (cf. Sect. 7.2.3 in Rubinstein and Colby). The latter is based on an increasing orientation of bonds in the direction of applied load when the strain increases. In a simple model description each bond is linked to an energy $fb \cos \alpha$, where α is the angle between the bond and the direction of the force f. Assuming in addition that the bonds are independent, it is easy to calculate the thermal average $\langle \cos \alpha \rangle$ (this calculation, even though in a different context, is carried out at the beginning of Sect. 5.3), i.e. $\langle \cos \alpha \rangle = \mathcal{L}(\beta f b)$, where $\mathcal{L}(x) = \coth(x) - 1/x$ is the **Langevin function** and $\beta = 1/(k_B T)$. Since $\langle \cos \alpha \rangle = R/Nb$, where R is the end-to-end distance of an already stretched chain consisting of N bonds of length b, we obtain $f(R)$ from the inverse Langevin function via

$$f(R) = (\beta b)^{-1} \mathcal{L}^{-1}(R/(Nb)). \tag{4.22}$$

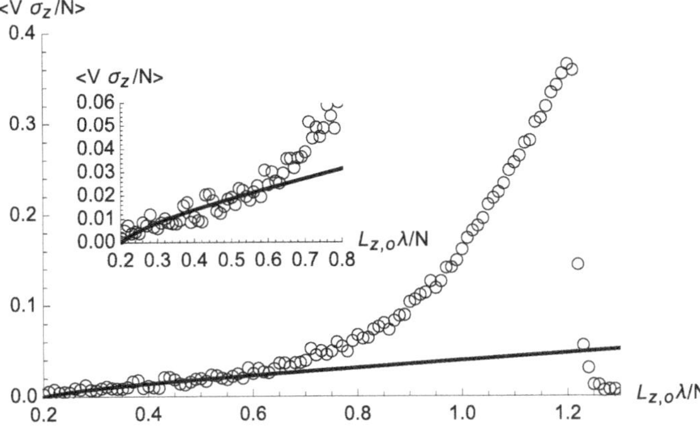

Fig. 4.5 Reduced stress versus reduced stretch for a Morse chain simulated using the Langevin Dynamics simulation technique. Data are obtained during a single run without extensive averaging. The z-dimension of the simulation box is $L_{z,o}\lambda$. Lines represent a fit to the data at low strain using (4.20). Here $N = 50$ and $L_{z,o} = 10 r_o$

This is a simple analytic expression relating the stretching force to the chain length in the regime of high elongation, whereas our result in Fig. 4.5 is a numerical one. However, the latter provides useful information regarding the breaking of the chain and its dependence on temperature and bond potential parameters.

Equations and Concepts from Fluid Mechanics:

The mass of fluid, whose uniform mass density is ρ, flowing through the surface of a volume V per unit time, is given by

$$\oint_V \rho \vec{v} \cdot d\vec{f} = \int_V \vec{\nabla} \cdot (\rho \vec{v}) \, dV, \tag{4.23}$$

where $\vec{v}$ is the fluid velocity within a volume element at position $\vec{r}$. Therefore the **mass continuity equation** becomes

$$\frac{\partial \rho}{\partial t} + \vec{\nabla} \cdot (\rho \vec{v}) = 0. \tag{4.24}$$

The total force acting on the volume is

$$-\oint_V P d\vec{f} = -\int_V \vec{\nabla} P \, dV, \tag{4.25}$$

where P is the pressure. Hence, the equation of motion of a liquid volume element is

$$\rho \frac{d\vec{v}}{dt} = -\vec{\nabla} P. \tag{4.26}$$

With $d/dt = \partial/\partial t + \vec{v} \cdot \vec{\nabla}$ the above equation becomes

$$\frac{\partial \vec{v}}{\partial t} + (\vec{v} \cdot \vec{\nabla}) \, \vec{v} = -\frac{1}{\rho} \vec{\nabla} P \tag{4.27}$$

and is called **Euler's equation**.

The momentum carried by the moving fluid per unit volume is $\rho \, \vec{v}$ and the attendant i-component of the momentum flux is

$$\frac{\partial}{\partial t} \rho \, v_i = \frac{\partial \rho}{\partial t} v_i + \rho \frac{\partial v_i}{\partial t}. \tag{4.28}$$

Using the continuity equation to replace $\partial \rho/\partial t$ and Euler's equation to replace $\partial v_i/\partial t$ we arrive at

$$\frac{\partial}{\partial t} \rho \, v_i = -\frac{\partial \Pi_{ik}}{\partial x_k}, \tag{4.29}$$

where

$$\Pi_{ik} = P \, \delta_{ik} + \rho v_i v_k. \tag{4.30}$$

In a **viscous fluid** we must add an extra contribution $-\tau'_{ik}$ to Π_{ik}, i.e.

$$\Pi_{ik} = P \, \delta_{ik} + \rho v_i v_k - \tau'_{ik}, \tag{4.31}$$

which describes the irreversible 'viscous' transfer of momentum in the fluid. In particular,

$$\tau_{ik} = -P \, \delta_{ik} + \tau'_{ik} \tag{4.32}$$

defines the **stress tensor** in this context.

We can deduce τ'_{ik} as follows: (i) Internal friction in the liquid implies that there must be velocity gradients $\partial v_i/\partial x_k$, and we assume that τ'_{ik} depends linearly on these gradients. (ii) When the liquid undergoes a uniform rotation we expect no friction, i.e. $\tau'_{ik} = 0$. The velocity due to this rotation is $\vec{v} = \vec{\omega} \times \vec{r}$ or, in terms of its components, $v_\alpha = \epsilon_{\alpha\beta\delta}\omega_\beta x_\delta$, where $\vec{\omega}$ is the angular velocity of the rotation and $\epsilon_{\alpha\beta\delta}$ are the components of the ϵ-**tensor**. Using the properties of the ϵ-tensor, we can quickly show that $\partial v_i/\partial x_k + \partial v_k/\partial x_i$ as well as $\partial v_l/\partial x_l \equiv \vec{\nabla} \cdot \vec{v}$ (summation

Table 4.2 Shear viscosity of selected substances

Substance	η (Pa s)
Air ($T = 300$ K)	$18.6 \cdot 10^{-6}$
Water ($T = 293$ K)	$1.002 \cdot 10^{-3}$
Olive oil ($T = 293$ K)	$80.8 \cdot 10^{-3}$
S-SBR M_w=276 [kg/mol]	about $1 \cdot 10^{11}$

convention!) are zero when rotated. Therefore we express τ'_{ik} as a linear combination of these two terms:

$$\tau'_{ik} = \eta \left(\frac{\partial v_i}{\partial x_k} + \frac{\partial v_k}{\partial x_i} - \frac{2}{3}\delta_{ik}\frac{\partial v_l}{\partial x_l} \right) + \eta'\delta_{ik}\frac{\partial v_l}{\partial x_l}. \tag{4.33}$$

We chose this combination so that the trace of the first term, multiplied by the **(shear) viscosity coefficient** η (cf. the selected examples in Table 4.2), is zero. The coefficient η' is the **second viscosity**.

Here we concentrate on incompressible liquids, which according to the continuity equation implies

$$\vec{\nabla} \cdot \vec{v} = 0 \tag{4.34}$$

and thus

$$\tau_{ik} = -P\delta_{ik} + \eta \left(\frac{\partial v_i}{\partial x_k} + \frac{\partial v_k}{\partial x_i} \right). \tag{4.35}$$

Note that the combination of (4.29), (4.31), (4.32), and (4.35) yields

$$\frac{\partial \vec{v}}{\partial t} + (\vec{v} \cdot \vec{\nabla})\vec{v} = -\frac{1}{\rho}\vec{\nabla}P + \frac{\eta}{\rho}\Delta\vec{v}, \tag{4.36}$$

where $\Delta = \vec{\nabla}^2$. This is the **Navier–Stokes equation** for an incompressible fluid. It will become important when we need to consider **hydrodynamic interactions** in polymer systems. We shall assume steady flow, i.e. $\partial\vec{v}/\partial t = 0$, and that the **Reynolds number** R, i.e.

$$R = \frac{\rho}{\eta}ul, \tag{4.37}$$

where l is a typical linear dimension of the body in the flow of the viscous liquid and u is the liquid's velocity far from the body, is small. Note that the quantity η/ρ is called **kinematic viscosity**. Small R implies that the term $(\vec{v} \cdot \vec{\nabla})\vec{v}$ can be neglected compared to $(\eta/\rho)\Delta\vec{v}$. This follows since the first term is on the order of u^2/l, whereas the second one is on the order of $(\eta/\rho)u/l^2 = (1/R)u^2/l$. Therefore (4.36) reduces to

$$\eta\Delta\vec{v} = \vec{\nabla}P. \tag{4.38}$$

If there is an additional external force density $\vec{\varphi}(\vec{r})$ present, this equation becomes

$$\eta\Delta\vec{v}(\vec{r}) - \vec{\nabla}P(\vec{r}) + \vec{\varphi}(\vec{r}) = 0. \tag{4.39}$$

Let us briefly digress and solve the above equation. This means that we do not have to do it later when we need the solution. We replace the components of $\vec{v}(\vec{r})$ and $\vec{\varphi}(\vec{r})$ as well as $P(\vec{r})$ by their Fourier transforms, i.e.

$$f(\vec{r}) = (2\pi)^{-3/2}\int d^3k f_k \exp[-i\vec{k}\cdot\vec{r}]. \tag{4.40}$$

The result is

$$-\eta k^2\vec{v}_k + i\vec{k}P_k + \vec{\varphi}_k = 0. \tag{4.41}$$

Scalar multiplication of this equation by $\vec{k}$ and using (4.34), i.e. $\vec{k}\cdot\vec{v}_k = 0$, yields

$$P_k = i\frac{\vec{k}\cdot\vec{\varphi}_k}{k^2}. \tag{4.42}$$

Inserting P_k into (4.41) we obtain the flow field in $\vec{k}$-space

$$\vec{v}_k = \frac{1}{\eta k^2}\left(\vec{\varphi}_k - \frac{\vec{k}(\vec{k}\cdot\vec{\varphi}_k)}{k^2}\right) \tag{4.43}$$

or in component form

$$v_{k,\alpha} = \frac{1}{\eta k^2}\left(\delta_{\alpha\beta} - e_{k,\alpha}e_{k,\beta}\right)\varphi_{k,\beta}, \tag{4.44}$$

where $\vec{e}_k = \vec{k}/k$. We find the flow field in $\vec{r}$-space using the **convolution theorem**, i.e.

$$v_\alpha(\vec{r}) = (2\pi)^{-3/2}\int d^3k \frac{1}{\eta k^2}\left(\delta_{\alpha\beta} - e_{k,\alpha}e_{k,\beta}\right)\varphi_{k,\beta}\exp[-i\vec{k}\cdot\vec{r}]$$

or

$$v_\alpha(\vec{r}) = \int d^3r' H_{\alpha\beta}(\vec{r}-\vec{r}\,')\varphi_\beta(\vec{r}\,'), \tag{4.45}$$

Fig. 4.6 Laminar flow
profile between two plates

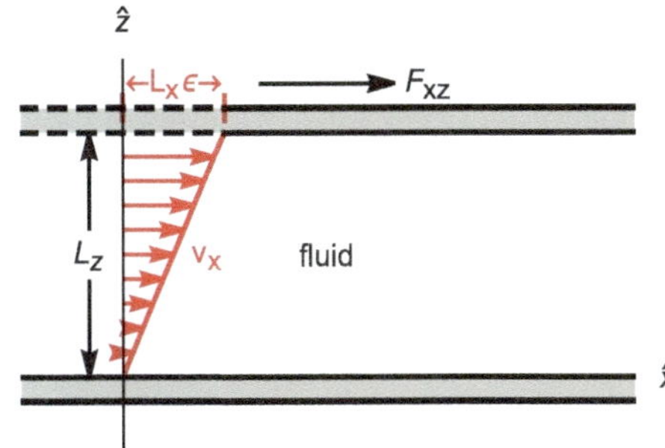

where

$$H_{\alpha\beta}(\vec{x}) = (2\pi)^{-3} \int d^3k \, \frac{1}{\eta k^2} \left(\delta_{\alpha\beta} - e_{k,\alpha} e_{k,\beta} \right) \exp[-i\vec{k} \cdot \vec{x}]. \tag{4.46}$$

The last step in our solution is the calculation of the integral in (4.46), which yields[2]

$$H_{\alpha\beta}(\vec{x}) = \frac{1}{8\pi\eta x} \left(\delta_{\alpha\beta} + e_{x,\alpha} e_{x,\beta} \right) \tag{4.47}$$

$(\vec{e}_x = \vec{x}/x)$. The quantities $H_{\alpha\beta}$ are the components of the **Oseen tensor**.

Now let's return to (4.35), which is of more immediate importance. Figure 4.6 depicts a fluid between two plates. The bottom plate is at rest while the upper plate moves in x-direction, causing a uniform velocity gradient in the fluid. The stationary profile is maintained by the force F_{xz} on the upper plate, i.e.

$$\tau_{xz} \propto \frac{\delta v_x}{\delta z} = \frac{L_x \epsilon}{\Delta t \, L_z} = \dot{\gamma}_{xz}, \tag{4.48}$$

where we use the same notation as in Fig. 3.8. Hence

$$\tau_{xz} = \eta \dot{\gamma}_{xz}. \tag{4.49}$$

---

[2] Note first that the change of variables $\vec{k}' = x\vec{k}$ immediately yields $H_{\alpha\beta}(\vec{x}) \propto x^{-1}$. The standard solution (cf. Doi and Edwards or Grosberg and Khokhlov) assumes the general form $H_{\alpha\beta}(\vec{x}) = A(x)\delta_{\alpha\beta} + B(x)e_{x,\alpha}e_{x,\beta}$. Taking the trace yields $H_{\alpha\alpha}(\vec{x}) = 3A(x) + B(x)$ or, from (4.46), $H_{\alpha\alpha}(\vec{x}) = (2\pi)^{-3} \int d^3k \, \frac{2}{\eta k^2} \exp[-i\vec{k} \cdot \vec{x}] = (2\pi\eta x)^{-1}$. We also have $e_{x,\alpha} H_{\alpha\beta}(\vec{x}) e_{x,\beta} = A(x) + B(x)$ and, again from (4.46), $e_{x,\alpha} H_{\alpha\beta}(\vec{x}) e_{x,\beta} = (2\pi)^{-3} \int d^3k \, \frac{1}{\eta k^2} \left(1 - (\vec{e}_x \cdot \vec{e}_k)^2\right) \exp[-i\vec{k} \cdot \vec{x}] = (4\pi\eta x)^{-1}$. Combination of the two equations for A and B yields $A = B = (8\pi\eta x)^{-1}$. Note also that the $\vec{k}$-integrals in $H_{\alpha\alpha}$ and $e_{x,\alpha} H_{\alpha\beta} e_{x,\beta}$ are easier compared to the one in (4.46). This is because $H_{\alpha\alpha}$ and $e_{x,\alpha} H_{\alpha\beta} e_{x,\beta}$ are invariant under rotations. Hence we can chose $\vec{x}$ as our z-axis, something we cannot do in (4.46), and use for instance spherical coordinates.

> A Phenomenological Model for Linear Viscoelasticity:

Equation (4.13) together with (4.49) furnish the basic elements in the phenomenological description of **linear viscoelasticity**. In the following we represent (4.13) in simplified form via

$$\sigma_G = G\,\gamma_G \tag{4.50}$$

and analogously (4.49) via

$$\sigma_\eta = \eta\,\dot{\gamma}_\eta. \tag{4.51}$$

A pictorial representation of the two equations is shown in Fig. 4.7. These basic elements can be combined into simple models describing linear viscoelastic material behavior. Three such models are depicted in Fig. 4.8. Here we consider model (a) only, the so-called Kelvin–Voigt (KV) model, mainly because it is well known to physicists.

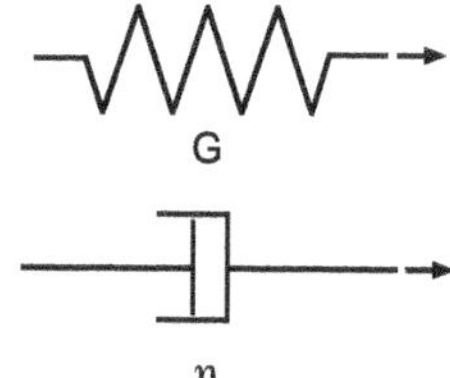

Fig. 4.7 Symbolic elements representing elastic and viscous contributions. The spring symbol stands for (4.13), whereas the dashpot symbol represents (4.49)

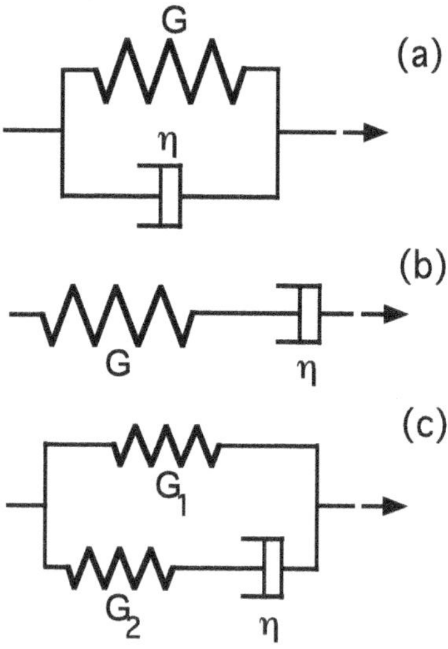

Fig. 4.8 a Kelvin–Voigt model; **b** Maxwell model; **c** Zener model

The mathematical form of the KV-model is

$$\sigma = \sigma_G + \sigma_\eta \qquad \gamma = \gamma_G = \gamma_\eta. \tag{4.52}$$

The quantity σ is the total stress and γ is the total strain. Due to the parallel arrangement of the two basic elements in the KV-model, σ is the sum of the stresses σ_G and σ_η contributed by the respective branches. The total strain γ, on the other hand, is identical to the individual strains in each branch, γ_G and γ_η, respectively. Using (4.50) and (4.51) we obtain for the KV-model

$$\sigma(t) = G\,\gamma(t) + \eta\,\dot{\gamma}(t). \tag{4.53}$$

Let's pause for a moment and compare this equation to the equation of motion of the one-dimensional damped harmonic oscillator driven by a time-dependent external force $f(t)$, i.e.

$$f(t) = k\,x(t) + \zeta\,\dot{x}(t) + m\,\ddot{x}(t). \tag{4.54}$$

Here $x(t)$ is the position of the oscillator relative to its position at rest, k is the spring constant, ζ is a friction constant, and m is the mass of the oscillator. The only difference between (4.53) and (4.54) is the absence of the acceleration term in (4.53). This means that the acceleration of the mass of a volume element within the sheared viscoelastic sample is not important. Instead the volume element moves according to the equilibrium of the forces acting on it.

We remember that the long-time solution of the oscillator subject to an oscillatory force possesses the same form as the force except for a phase shift. Here we try an analogous ansatz in the case of (4.53), i.e.

$$\sigma(t) = \sigma_o \sin(\omega t + \delta) \quad \text{and} \quad \gamma(t) = \gamma_o \sin(\omega t). \tag{4.55}$$

We may be surprised that the stress ansatz includes the phase shift δ and not the strain. Figure 4.9 depicts the type of **rheometer** frequently used in shear experiments when the sample is an elastomer. The specimen, a disc-shaped piece of rubber in this case, is clamped between two dies. One of the dies imposes a rotary oscillation upon the specimen. The corresponding angle at time t is $\varphi(t)$. The other die keeps the opposite side of the sample stationary during the oscillation, which shears the sample. Keeping the opposite side of the sample fixed requires the torque $N_\varphi(t)$. An interesting feature of this design is the biconical shape of the die assembly (the sample is thick at its edge but becomes thinner towards its center), which ensures that the strain inside the sample is uniform. The protrusions on the sample surface are necessary for a tight grip of the die on the sample. Note that $\gamma(t)$ and $\sigma(t)$ are related to $\varphi(t)$ and $N_\varphi(t)$ via

$$\gamma(t) = \frac{\varphi(t)}{\phi} \quad \text{and} \quad \sigma(t) = \frac{3N_\varphi(t)}{2\pi R^3}. \tag{4.56}$$

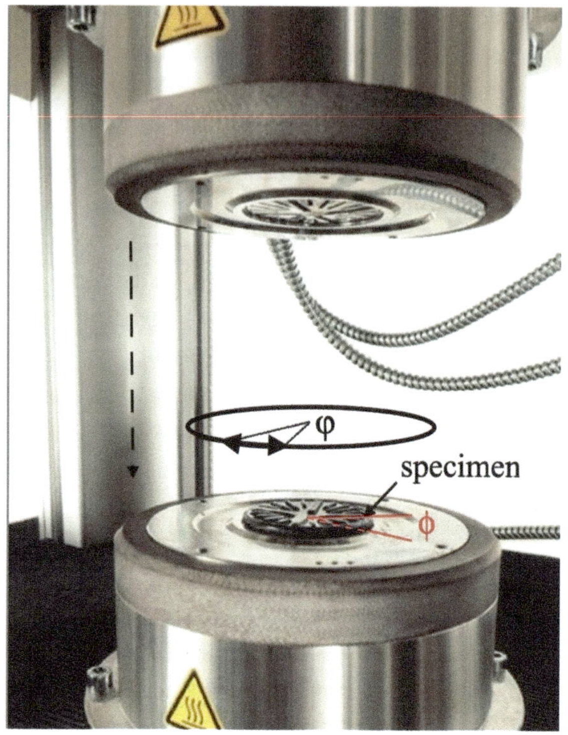

Fig. 4.9 A MonTech® Moving Die Rheometer with biconical die assembly. The biconical rubber sample, whose radius is R, is thin at its center and gets wider with increasing radial distance from the center. The corresponding angle is ϕ

A typical value of ϕ is $7°$. This means that a rotation φ of $1°$ corresponds to a strain of $\varphi/\phi \approx 14\%$.

Inserting the ansatz (4.55) into (4.53) and using the identity

$$\sin(\omega t + \delta) = \cos(\delta)\sin(\omega t) + \sin(\delta)\cos(\omega t) \tag{4.57}$$

yields

$$G' \equiv \frac{\sigma_o}{\gamma_o}\cos\delta = G \tag{4.58}$$

$$G'' \equiv \frac{\sigma_o}{\gamma_o}\sin\delta = \eta\omega \tag{4.59}$$

$$\tan\delta = \frac{G''}{G'} = \tau_{KV}\omega \quad (\tau_{KV} = \eta/G). \tag{4.60}$$

Here G' is the **storage modulus**, G'' is the **loss modulus**, and their ratio, $\tan \delta$, is the **loss tangent**. The quantity τ_{KV} is a relaxation time. We see this if we set the stress in (4.53) equal to zero. Separation of variables then leads to

$$\gamma(t) = \gamma(0) \exp[-t/\tau_{KV}], \tag{4.61}$$

which describes the strain relaxation after the stress is 'turned off'.

However, let's discuss G' and G'' in more detail. First of, why are they important? Their measurement yields information on the dependence of the dynamic mechanical properties of a viscoelastic material on frequency, temperature, and composition. From this information a material developer of, for instance, tire tread rubber can estimate a new tire's performance during breaking or what its rolling resistance will be compared to some other tire. G' and G'' are also important in the food industry, just to give you an impression of the range of their application, since they affect the processing and even determine how food feels when you eat it (e.g., Chaps. 4–6 in [6]).

G' and G'' are difficult quantities—both theoretically and experimentally. Our results in (4.58), i.e. $G' = G$ and $G'' = \eta \omega$, are specific to the KV-model. Had we used the Maxwell model in Fig. 4.8 the result would have been different. The Zener model, we would find, is a combination of both the KV- and the Maxwell model. The problem with these phenomenological models is that they do not connect well to the molecular scale. This means that we have to do the measurement first and decide only afterwards which phenomenological model yields the best qualitative description of the experimental data. Nevertheless, phenomenological models of viscoelasticity allow a rough understanding of the interplay between elasticity and viscosity—and they are simple, which makes them sufficiently useful.

If $G' = G$ and $G'' = \eta \omega$ is a result specific to the KV-model, how general are the definitions $G' \equiv \frac{\sigma_o}{\gamma_o} \cos \delta$ and $G'' \equiv \frac{\sigma_o}{\gamma_o} \sin \delta$? These definitions are general as long as the ansatz (4.55) works, i.e. equations into which we insert the ansatz must be linear in $\gamma(t)$ and $\sigma(t)$. Rheometer like the one depicted in Fig. 4.9 are sold with software which analyses **stress-strain curves**. For a linear viscoelastic sample, i.e. the ansatz works, the experimental result will look like the elliptical closed curve shown in Fig. 4.10. Each closed curve corresponds to a complete cycle of $\gamma(t)$. The enclosed area is the dissipated energy during the cycle. Let's calculate this energy:

$$w = \oint \sigma \, d\gamma = \int_0^{2\pi/\omega} \sigma \, \dot{\gamma} \, dt. \tag{4.62}$$

Inserting (4.55) yields

$$w = \pi \gamma_o \sigma_o \sin \delta = \pi G'' \gamma_o^2, \tag{4.63}$$

i.e. only G'' does contribute to w. Hence its name—loss modulus. Note that w is the dissipated energy per unit volume. The three stress-strain curves in Fig. 4.10 therefore describe no loss $w(\delta = 0) = 0$, intermediate loss $w(\delta = \pi/10)$, maximum loss

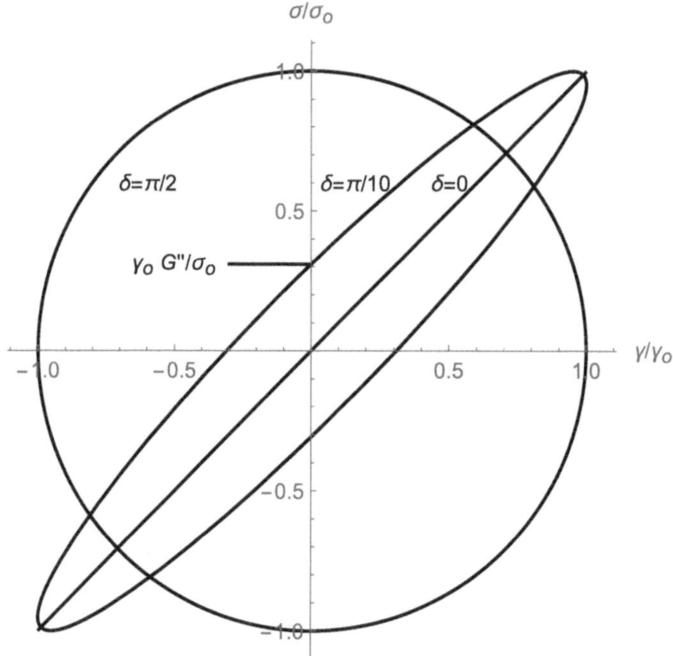

Fig. 4.10 Stress-strain cycles for three different phase shifts

$w(\delta = \pi/2)$. In other words, the rheometer's software will try and fit the recorded stress-strain data using (4.55), which yields δ and σ_o and, employing the definitions $G' \equiv \frac{\sigma_o}{\gamma_o} \cos \delta$ and $G'' \equiv \frac{\sigma_o}{\gamma_o} \sin \delta$, G' and G''. However, the experimental stress-strain curves may be distorted (e.g., when the shear modulus depends on the strain amplitude for various reasons). A schematic example of a distorted stress-strain/torque-rotation angle-cycle is depicted in Fig. 4.11. Despite this the software still attempts to find the best fit based on the linear theory. In fact, many experimental results for G' and G'' in the literature have been obtained in this fashion (which, if one is aware of this procedure, still yields useful information[3]).

Remark: Equation (4.62) is completely general and independent of whether the viscoelastic behavior is linear or not. Equation (4.63) therefore could be used as an independent definition of G'', i.e. G'' would solely result from the enclosed area of the stress-strain curve.

---

[3] This is under the assumption that the structure of the material is responsible for the distortion and not, for instance, slippage in the interface between die and sample, which may occur at large strain amplitudes.

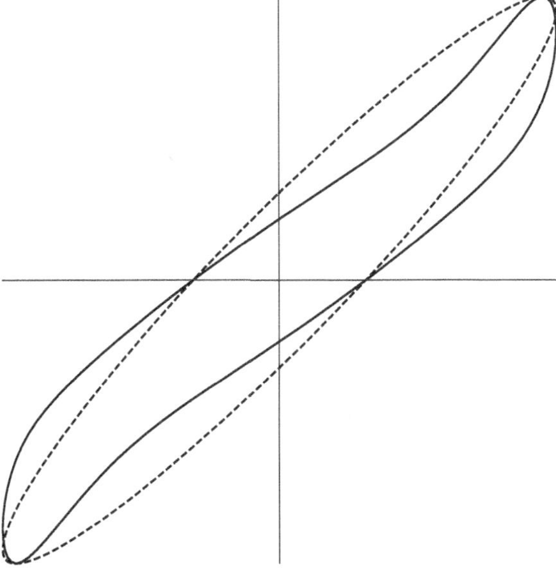

Fig. 4.11 Schematic of a distorted stress-strain cycle compared to its elliptical approximation

The real ansatz (4.55) can be replaced by its complex version:

$$\tilde{\sigma} = \sigma_o e^{i(\omega t + \delta)} \qquad \tilde{\gamma} = \gamma_o e^{i\omega t}. \tag{4.64}$$

It is common practice to define the **complex modulus** $G^* = \tilde{\sigma}/\tilde{u}$. Insertion of (4.64) into (4.53) yields

$$G^* = G' + iG'', \tag{4.65}$$

i.e. G' and G'' are the real and imaginary parts of the complex modulus. The inverse of the complex modulus G^* is called **compliance** J^*, i.e.

$$J^* = \frac{1}{G^*}. \tag{4.66}$$

The compliance is a measure for how easy it is to deform the sample. Note

$$J' = \frac{G'}{|G^*|^2} \quad \text{and} \quad J'' = \frac{G''}{|G^*|^2}, \tag{4.67}$$

where $|G^*|^2 = G'^2 + G''^2$.

Finally, let us look at some experimental results. The solid symbols in Fig. 4.12 depict the shear storage modulus of S-SBR and the open symbols show the corresponding shear loss modulus. Both quantities are plotted versus frequency f. You

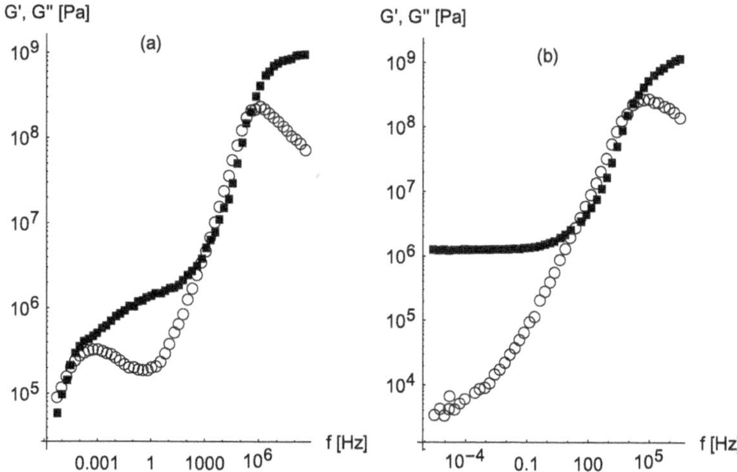

Fig. 4.12 Comparison of the master curves for S-SBR samples without chemical cross-links (a: 0 phr) and with chemical cross-links (b: 4 phr sulfur). Solid symbols: G'; open symbols: G''. The reference temperature is 23 °C. 0 or 4 phr (parts per hundred rubber by weight) is the amount of sulfur added in order to create chemical cross-links during vulcanization. All data in this figure are from Fig. 2.99 in [7]

may wonder how a machine like the one shown in Fig. 4.9 is capable of up to 10^8 Hz. Well, it is not. The rheometer acquires data at much lower frequencies in a comparatively narrow frequency window—perhaps 0.01–1 Hz. Subsequently, using what is called **time-temperature superposition**, which we discuss below, **master curves** are constructed—like the ones shown in Fig. 4.12. The meaning of the master curves is that if we had an instrument capable of covering the indicated frequency window, then it would yield this result at 23 °C.

Let us concentrate on panel (b) first. Below 1 Hz we obtain the result predicted by the KV-model in (4.58) and (4.59). The storage modulus is constant and the loss modulus decreases with decreasing frequency, albeit with a power less than one. Above 1 Hz, however, G' rises and appears to level off eventually. G'' on the other hand exhibits a maximum. In panel (a) the behavior of G' and G'' above 1 Hz is pretty much unaltered in comparison to panel (b). But below 1 Hz the storage modulus decreases with variable degree of steepness and G'' exhibits another maximum. The main difference between the two panels is the amount of cross-linking. The results in panel (a) are for a polymer, which is not chemically cross-linked. Hence, when the shear frequency is low, the polymer chains possess much more mobility as compared to panel (b). Finally, Fig. 4.13 shows that a constant storage modulus is not just due to chemical cross-linking. In this case the so-called **plateau modulus** develops by increasing the molecular weight. As the polymer chains get longer we expect additional **entanglement**—again an obstacle to the mobility of certain modes of motion the chains do otherwise possess.

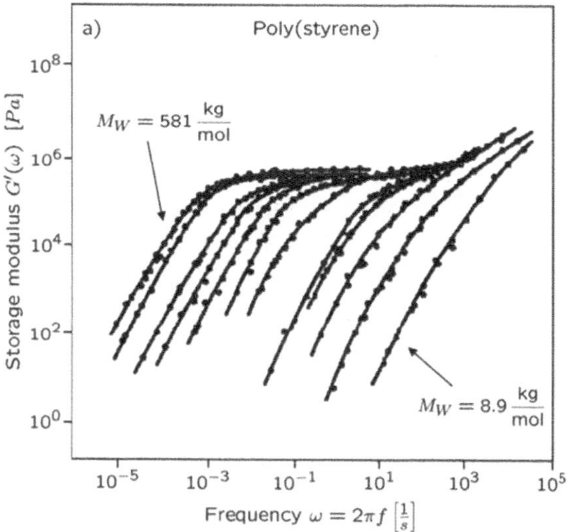

Fig. 4.13 Frequency dependence of the storage modulus of polystyrene in the molten state as function of molecular weight. This figure is copied from [7] (Fig. 2.82). Original source: [8]. Reprinted with permission from C. Wrana and the American Chemical Society

Overall, this should be obvious at this point, our understanding of G' and G'' as functions of frequency is insufficient and we must aim for a better and more comprehensive understanding—preferably on the molecular level.

Time-Temperature Superposition:

Figure 4.14 shows measurements of both the storage (top) and the loss modulus (bottom) of a SBR sample. On the left hand side of this figure the original data are depicted. They consist of measurements at different temperatures in a frequency range between roughly 0.01–1 Hz.

The arrows indicate the next step in the construction of a master curve. The red data points obtained at $T_r = 20\,^{\circ}$C—the arbitrarily chosen reference temperature—are left untouched. All other data sets, obtained at temperatures T_i, are shifted horizontally along the frequency axis. This shift is towards the right, i.e. towards higher frequencies, when $T_i < T_r$. Otherwise, $T_i > T_r$, it is towards the left, i.e. towards lower frequencies. The amount of shifting is controlled by the condition that the shifted data sets form a smooth continuous curve. The claim is that this curve yields the same G' or the same G'', at T_r, which one obtains with an instrument actually capable of scanning the entire frequency range from 10^{-3} Hz to 10^{11} Hz in Fig. 4.14. The final result looks good, but is there a justification for this procedure?

Looking at our only expressions for G' and G'' thus far, i.e. (4.58) and (4.59), we notice that $\omega = 2\pi f$ appears multiplied with the viscosity η, which is strongly dependent on temperature T. In other words, if frequency enters into a dynamical expression g via the product $\eta(T)\omega \equiv y$, i.e. $g = g(y)$, we can obtain the value $g(y)$ using either the product of $\eta(T')$ with ω' or η with ω as long as $y = \eta(T')\omega' = \eta(T)\omega$. Hence, if a particular dynamical feature (a particular measured value for G' or G''

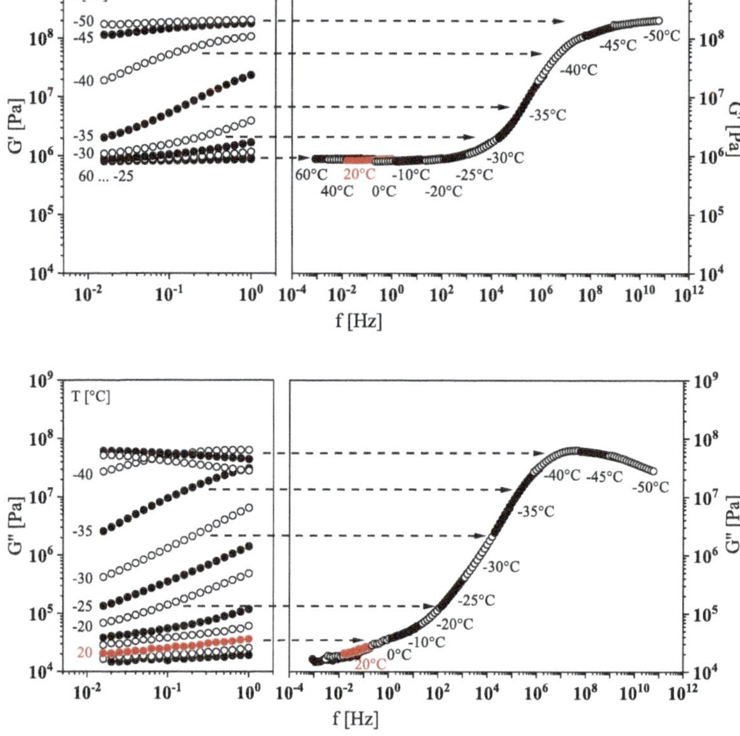

Fig. 4.14 Illustration of time-temperature superposition. Reprinted with permission from [9]

in the original data sets depicted in Fig. 4.14) appears at the frequency ω' when the temperature is T', then the very same feature will appear at the frequency ω when the temperature is T. Hence,

$$\omega = \frac{\eta(T')}{\eta(T)}\omega' \qquad (4.68)$$

converts the original frequencies into the shifted ones.

Let us find out what the η-ratio is in terms of explicit temperature. The temperature dependence of η is described empirically by the **Dolittle relation**

$$\eta(T) \propto \exp[A/v_f(T)]. \qquad (4.69)$$

The quantity A is a constant and $v_f(T)$ is the **free volume**. The free volume is illustrated in Fig. 4.15 for a spherical particle moving freely within a larger and likewise spherical cavity. The volume indicated by the dashed circle in the figure is the volume accessible to the center of mass of the particle, which is its free volume

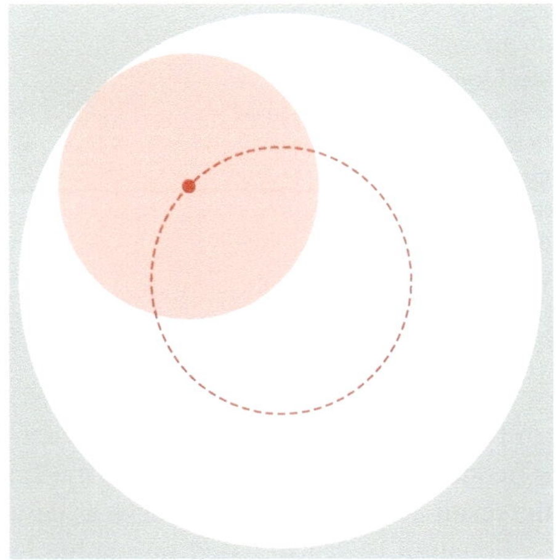

Fig. 4.15 Free volume illustration. The particle is shown in light red

in this example. Mathematically the free volume of a particle within a potential $u(\vec{r})$ is given by

$$v_f = \int_V d^3r \, \exp[-\beta(u(\vec{r}) - u(0))], \tag{4.70}$$

where $\beta = (k_B T)^{-1}$ and $u(0)$ is the potential minimum within the cavity. Clearly, as the shape of the particle as well as the shape of the cavity become more complicated so does v_f. Nevertheless, to leading order we may express the temperature dependence of v_f via the linear equation

$$v_f(T) = v_f(T_r) + \alpha(T - T_r). \tag{4.71}$$

Here α is a constant and T_r is an arbitrary reference temperature.

Inserting (4.71) into (4.69) we obtain for the ratio $\eta(T)/\eta(T_r)$:

$$a_T \equiv \frac{\eta(T)}{\eta(T_r)} = \exp\left[-\frac{A}{v_f(T_r)} \frac{T - T_r}{v_f(T_r)/\alpha + T - T_r}\right]. \tag{4.72}$$

This is the **WLF equation**, named after Williams, Landel, and Ferry, who first applied it to polymer melts [10–12]. The quantity a_T is also called the **shift factor**. Hence, the shifted frequencies in Fig. 4.14 are obtained from their reference or original values ω_r via

$$\omega = \frac{1}{a_T}\omega_r. \tag{4.73}$$

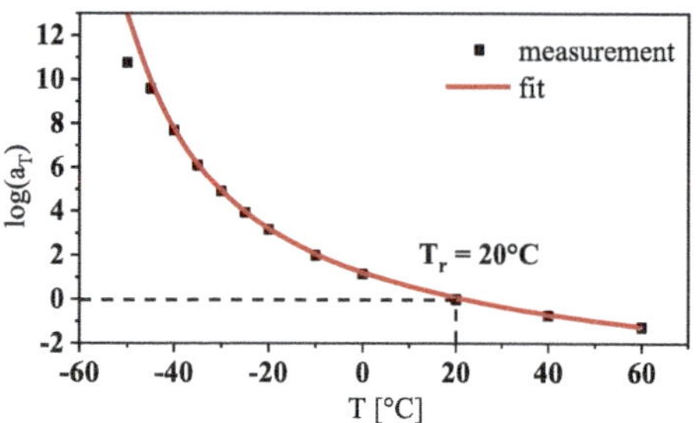

Fig. 4.16 Shift factor based on the data in Fig. 4.14. Each solid square corresponds to a series of data points obtained in the original frequency intervall (here 0.01–1 Hz) at the respective temperature. Reprinted with permission from [9]

The shift factor belonging to the master curves in Fig. 4.14 is depicted in Fig. 4.16. a_T contains the two parameters $A/v_f(T_r)$ and $v_f(T_r)/\alpha$, which are adjusted until the result is a smooth master curve as described above.

Equation (4.72) is based on qualitative arguments and rough approximations. Despite this, time-temperature superposition works extremely well in pure polymer systems and is frequently used. In fact, 'mastering' is not restricted to G' and G'', but is applied to other frequency/temperature-dependent quantities as well. Before we leave this section, let us briefly study an important example in the context of rubber **sliding friction**.

In school or in undergraduate physics courses the coefficient of sliding friction μ is a material or interface dependent number. The top panel in Fig. 4.17 depicts μ for various rubbers sliding on glass over a wide range of sliding velocities v. This master curve for μ is obtained via the same procedure used to obtain the master curves for the **dynamic moduli**. Measurements of $\mu(v)$ in a comparatively narrow velocity interval are carried out at different temperatures. Subsequently the individual measurements are combined into a smooth master curve via horizontal shifting.

The first aspect we notice is that μ is not a constant at all. It starts low when the sliding velocity is small and it ends low when the sliding velocity is large. Inbetween it exhibits a pronounced maximum. The bottom panel in Fig. 4.17 shows the corresponding loss moduli versus frequency f. Apparently there is a close correspondence between the maximum of μ and the maximum of G'' (Note that similar maxima can be seen in Figs. (4.12) and (4.14) as well.). Can we understand why there should be this connection between μ and G''?

The sketch in Fig. 4.18 depicts the rubber sample sliding on a surface. The surface's roughness leads to the deformation of the rubber, which in turn leads to a viscoelastic response indicated by the two pictures of the KV model. Assuming linearity, the energy (per volume rubber) dissipated during sliding is given by (4.63). We can also express the dissipated energy as the product of the friction force, which itself is proportional to μ, times a certain sliding distance s. Combination of the two observations suggests

$$\mu(v) \propto G''(f), \tag{4.74}$$

Fig. 4.17 Comparison between the friction coefficient master curves and the loss modulus master curves for five not chemically cross-linked rubbers on dry glass. The reference temperature is 20 °C, i.e. the shift factor $a_T = 1$ at this temperature. For other temperatures we need to know the corresponding a_T-value in order to calculate the attendant frequency. Reprinted with permission from [13]

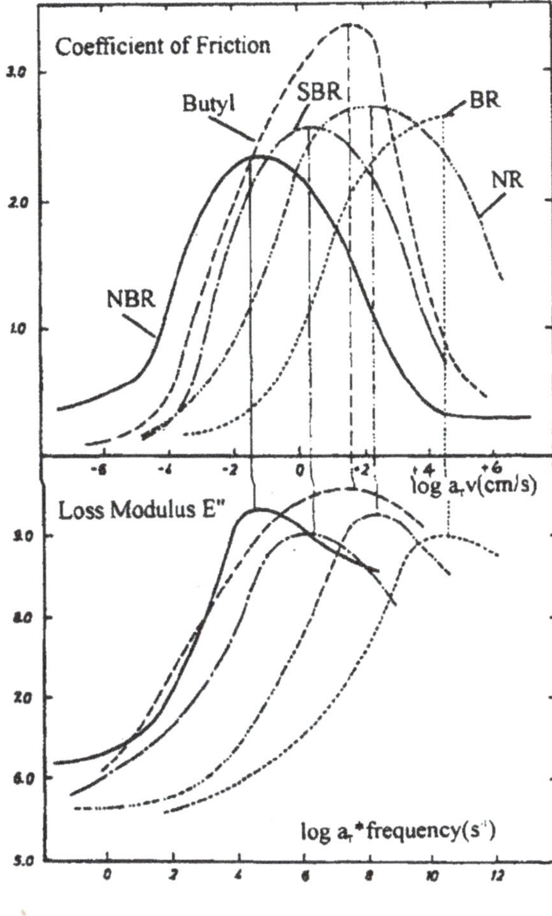

Fig. 4.18 Cartoon of a piece of rubber sliding on a rough surface with the velocity $\vec{v}$

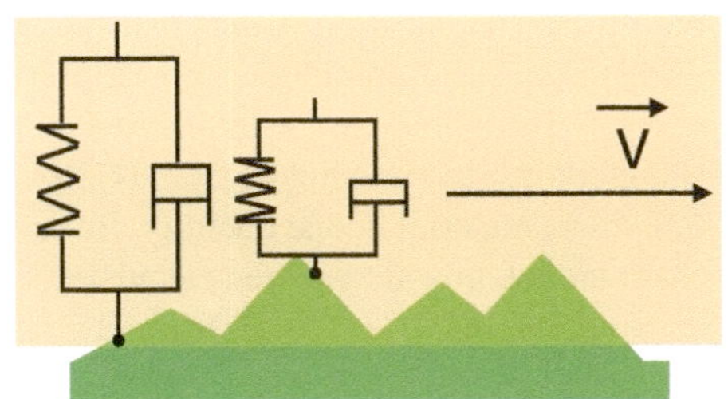

where $f = f(v)$. If we simply assume that s is the distance between two (identical) protrusions deforming the rubber then $v = sf$, which explains why we find velocity on the x-axis of the μ-master curve and frequency on the x-axis of the G''-master curve. However, it is important to note that we assume that the only contribution to friction is **hysteresis friction**. This is never quite the case, since we always have to include **adhesion** and **mechanical interdigitation** as well. In addition, the complex morphology of the interface between rubber and the solid must be taken into account. Hence, it is not surprising that friction of polymer materials is the subject of active research (e.g., [14]).

The Shear Relaxation Modulus:

Our discussion of linear viscoelasticity thus far is based on the ad hoc combination of σ_G and σ_η, i.e. the combination of (4.50) and (4.51). The specific time or frequency dependence of the measurable quantities, e.g., the storage modulus G' or the loss modulus G'', varies depending on the particular combination of σ_G and σ_η we choose to study. In addition, we only know how the calculate G' and G'' based on phenomenological models like the ones depicted in Fig. 4.8. All in all this is not a satisfactory situation.

In the next section we study the dynamics of polymer chain models. Since these models possess clear physical meaning compared to the phenomenological models, we want to be able to calculate their G' and G''. But how do we calculate the **dynamic moduli** when we do not have an equation like (4.53) for the KV-model?

The equation

$$\sigma(t) = \int_{-\infty}^{t} G(t - t')\, d\gamma(t') \tag{4.75}$$

is a generalization of (4.53) and others like it. G is called the **shear relaxation modulus**. The equation means that the shear stress at time t is accumulated from many small shear amplitude steps at previous times t'. In essence, the contribution of a small amplitude to the final stress depends on the time interval $t - t'$. $G(t - t')$ is a modulus, but it is not G from (4.50). It combines the viscoelastic nature of the (polymer) material in one quantity. But even if we have this $G(t - t')$ for a specific chain model, how do we then calculate G' and G''?

Note first that $d\gamma(t) = \dot{\gamma}(t)dt$ and thus (4.75) becomes

$$\sigma(t) = \int_{-\infty}^{t} G(t - t')\, \dot{\gamma}(t')dt'. \tag{4.76}$$

If for instance the strain increases instantaneously from zero to its maximum value at $t = 0$, i.e.

$$\gamma(t) = \begin{cases} 0 & t < 0 \\ \gamma_o & t > 0 \end{cases}, \tag{4.77}$$

then (4.76) yields

$$\sigma(t) = G(t)\gamma_o. \tag{4.78}$$

Here we have used $\dot{\gamma}(t) = \gamma_o \delta(t)$. This means that if we can realize (4.77) experimentally, then we can obtain $G(t)$ by observing the decay of the shear stress.

Of particular interest is a sinusoidal shear strain, i.e.

$$\gamma(t) = \gamma_o \sin(\omega t) = \gamma_o \, \text{Im}(e^{i\omega t}) \tag{4.79}$$

(cf. (4.55)), since we already have discussed this type of shearing. Using (4.55) in conjunction with (4.57) and $G' = \frac{\sigma_o}{\gamma_o} \cos\delta$ as well as $G'' = \frac{\sigma_o}{\gamma_o} \sin\delta$ (cf. (4.58) and (4.59)) we obtain

$$\sigma(t) = G'(\omega)\gamma_o \sin(\omega t) + G''(\omega)\gamma_o \cos(\omega t) = \gamma_o \, \text{Im}\left(G^*(\omega)e^{i\omega t}\right). \tag{4.80}$$

We can now show the validity of

$$G^*(\omega) = i\omega \int_0^\infty dt \, G(t) \, e^{-i\omega t} \tag{4.81}$$

by inserting (4.81) into (4.80).[4] From (4.81) we immediately find the desired equations for the storage and the loss modulus in terms of $G(t)$, i.e.

$$G'(\omega) = \omega \int_0^\infty dt \, G(t) \, \sin(\omega t) \tag{4.83}$$

and

$$G''(\omega) = \omega \int_0^\infty dt \, G(t) \, \cos(\omega t). \tag{4.84}$$

Application examples of these two useful formulas will be discussed in the next section.

---

[4]

$$\sigma(t) = \gamma_o \, \text{Im}\left(i\omega \int_0^\infty dt' \, G(t') \, e^{i\omega(t-t')}\right) \overset{t-t'=t''}{=} \gamma_o \int_{-\infty}^t dt'' G(t-t'') \, \text{Im}\left(i\omega e^{i\omega t''}\right) \tag{4.82}$$

Working out Im(...) we find that the result is again (4.76).

4.2   Single Chain Dynamics

In this section we discuss polymer dynamics based on a single chain model and its refinements. And we shall obtain $G(t)$ based on this model.

Preliminaries:

Imagine a sphere of mass m and radius R moving in a liquid of small molecules possessing the mass density ρ and the viscosity η. The equation of motion of (the center of mass of) this sphere is

$$m\,\ddot{\vec{r}} = -\zeta\,\dot{\vec{r}}. \tag{4.85}$$

The right hand side is **Stoke's friction** force, i.e. $\zeta = 6\pi\eta R$. This may be expressed as

$$\dot{\vec{v}} = -\frac{1}{\tau}\vec{v}, \tag{4.86}$$

where $\vec{v} = \dot{\vec{r}}$ and

$$\tau = \frac{m}{\zeta} = \frac{2}{9}\frac{\rho R^2}{\eta}. \tag{4.87}$$

Here we assume that the sphere's mass density is also ρ. Equation (4.86) tells us that the initial velocity of the sphere decays exponentially with a relaxation time τ, i.e.

$$\vec{v}(t) = \vec{v}(0)\exp[-t/\tau]. \tag{4.88}$$

Assuming the liquid is water and using $R = 10\,\mathrm{nm}$ we find $\tau \approx 2 \cdot 10^{-11}\,\mathrm{s}$. This is short compared to (most) relaxation times in polymer dynamics.

It is possible and useful to tie the friction coefficient ζ to the diffusion coefficient D in the **Einstein law of diffusion** (2.6). Our starting point is

$$6Dt = \langle \Delta\vec{r}^{\,2} \rangle = \int_o^t dt' \int_o^t dt'' \langle \vec{v}(t') \cdot \vec{v}(t'') \rangle \tag{4.89}$$

or

$$6D = \frac{d}{dt}\langle \Delta \vec{r}^{\,2}\rangle = 2\langle \vec{v}(t) \cdot [\vec{r}(t) - \vec{r}(0)]\rangle \tag{4.90}$$

$$= 2\langle \vec{v}(0) \cdot [\vec{r}(0) - \vec{r}(-t)]\rangle$$

$$= 2\int_{-t}^{0} dt'\langle \vec{v}(0) \cdot \vec{v}(t')\rangle$$

$$= 2\int_{0}^{t} dt'\langle \vec{v}(0) \cdot \vec{v}(t')\rangle$$

$$= 2\langle \vec{v}(0)^2\rangle \int_{0}^{t} dt' \exp[-t'/\tau]$$

$$= 2\langle \vec{v}(0)^2\rangle \tau (1 - \exp[-t/\tau]).$$

If we now use $\langle \vec{v}(0)^2\rangle = 3k_B T/m$ (note that $m\langle \vec{v}(0)^2\rangle/2$ is the average kinetic energy of the sphere) and assume $t \gg \tau$ we obtain

$$D = \frac{k_B T}{\zeta}. \tag{4.91}$$

Next we want to study the sustained **Brownian motion** of our sphere. We do this by adding a random force $\vec{Z}$ with $\langle \vec{Z}\rangle = 0$ to (4.85), i.e.

$$\dot{\vec{v}} = -\frac{1}{\tau}\vec{v} + \frac{1}{m}\vec{Z}. \tag{4.92}$$

The solution of (4.92) is

$$\vec{v}(t) = \vec{v}(0)e^{-t/\tau} + \frac{1}{m}e^{-t/\tau}\int_{0}^{t} dt'\vec{Z}(t')e^{t'/\tau}. \tag{4.93}$$

This can be checked by inserting (4.93) into (4.92).

Here we are interested in the auto-correlation function $\langle Z(t')Z(t'')\rangle$, since we shall need it when we discuss the dynamics of the models of Rouse and Zimm. We find this auto-correlation function by working out the quantity $\langle \vec{v}(t)^2\rangle$ using (4.93). As before $\langle \vec{v}(t)^2\rangle = 3k_B T/m$ and thus

$$\frac{3k_B T}{m} = \langle \vec{v}(t)^2\rangle = \frac{1}{m^2}e^{-2t/\tau}\int_{0}^{t} dt' \int_{0}^{t} dt''\langle \vec{Z}(t') \cdot \vec{Z}(t'')\rangle e^{(t'+t'')/\tau}. \tag{4.94}$$

On the right hand side we have used $t \gg \tau$ and $\langle \vec{Z} \cdot \vec{v}\rangle = 0$. A guess which solves (4.94), as we show via straightforward insertion, is

$$\langle \vec{Z}(t') \cdot \vec{Z}(t'')\rangle = 6\zeta k_B T \delta(t' - t''). \tag{4.95}$$

If we extend (4.92) by a force derived from a potential U, i.e. $-\vec{\nabla}U$, we obtain the **Langevin equation** of motion:

$$m\dot{\vec{v}} = -\zeta\vec{v} - \vec{\nabla}U + \vec{Z}. \tag{4.96}$$

We conclude our preliminary considerations with a specific example—the Brownian motion of a one-dimensional harmonic oscillator, i.e.

$$U(x) = \frac{1}{2}k\,x^2, \tag{4.97}$$

where k is the force constant. If we are interested in times significantly greater than τ, we expect that the inertia term in the Langevin equation is not important (cf. our numerical example in the context of (4.88)). Hence it is sufficient to study the balance of forces on the right hand side of (4.96), i.e. in one dimension

$$\frac{d}{dt}x(t) = -\frac{k}{\zeta}\,x(t) + \frac{1}{\zeta}Z_x(t). \tag{4.98}$$

Here

$$\tau' = \frac{\zeta}{k} \tag{4.99}$$

is another relaxation time different from τ! The solution, as we can easily check, is

$$x(t) = \frac{1}{\zeta}\int_{-\infty}^{t} dt'\, e^{-(t-t')/\tau'}\, Z_x(t'). \tag{4.100}$$

From this we obtain the position auto-correlation function

$$\langle x(0)x(t)\rangle = \frac{1}{\zeta^2}\int_{-\infty}^{t} dt'\int_{-\infty}^{0} dt''\, e^{-(t-t'-t'')/\tau'}\,\langle Z_x(t')Z_x(t'')\rangle. \tag{4.101}$$

Assuming that the auto-correlation function $\langle \vec{Z}(t')\cdot\vec{Z}(t'')\rangle$ is not affected by U, we can insert it from (4.95). Hence,

$$\langle x(0)x(t)\rangle = \frac{k_B T}{k}\,\exp[-t/\tau']. \tag{4.102}$$

This is a result we shall return to when we study the Rouse chain. Another function we shall need is the root-mean-square displacement $\langle (x(t) - x(0))^2\rangle$ of the oscillator, i.e.

$$\langle (x(t) - x(0))^2 \rangle = \langle x(t)^2 \rangle + \langle x(0)^2 \rangle - 2\langle x(0)x(t) \rangle$$

$$= 2\left(\langle x(0)^2 \rangle - \langle x(0)x(t) \rangle \right)$$

$$= 2\frac{k_B T}{k}\left(1 - e^{-t/\tau'} \right) \qquad (4.103)$$

$$\approx \begin{cases} 2Dt & t \ll \tau' \\ 2\frac{k_B T}{k} & t \gg \tau' \end{cases}.$$

This means that at short times the root-mean-square displacement of the oscillator is diffusion controlled, whereas for times significantly longer than τ' it is potential controlled. Note that the limit $t \ll \tau'$ really means $\tau \ll t \ll \tau'$.

The Models of Rouse and Zimm:

Figure 4.19 illustrates the **Rouse model** [15]. Its construction begins with a freely jointed chain of Kuhn segments $\vec{b}_i$. The Kuhn segments are replaced with harmonic springs and the junctions of adjacent Kuhn segments become beads possessing masses m_i. The force constants of the springs are

$$k = \frac{3k_B T}{b^2}. \qquad (4.104)$$

This assumes that the springs themselves are entropy elastic (cf. (2.68)).

The equation of motion for bead i in the chain, analogous to (4.98) for the one-dimensional oscillator, is given by

$$\frac{d}{dt}\vec{r}(i, t) = \frac{k}{\zeta}\frac{d^2\vec{r}(i, t)}{di^2} + \frac{1}{\zeta}\vec{Z}(i, t). \qquad (4.105)$$

But how do we explain the replacement of $-x(t)$ by $d^2\vec{r}(i, t)/di^2$? Note that for any bead i possessing a left and a right neighbor

$$-\vec{\nabla}_i U = -\left[k(\vec{r}_i - \vec{r}_{i+1}) - k(\vec{r}_{i-1} - \vec{r}_i) \right] \approx k\frac{d^2\vec{r}(i, t)}{di^2}. \qquad (4.106)$$

We can also use this equation for the first bead $i = 0$ and the last bead $i = N$, when we provide them with 'artifical' neighbors located at $\vec{r}_{-1} = \vec{r}_0$ and $\vec{r}_{N+1} = \vec{r}_N$, respectively. This implies the boundary conditions

$$\frac{d\vec{r}(i, t)}{di} = 0 \quad \text{for} \quad i = 0, N. \qquad (4.107)$$

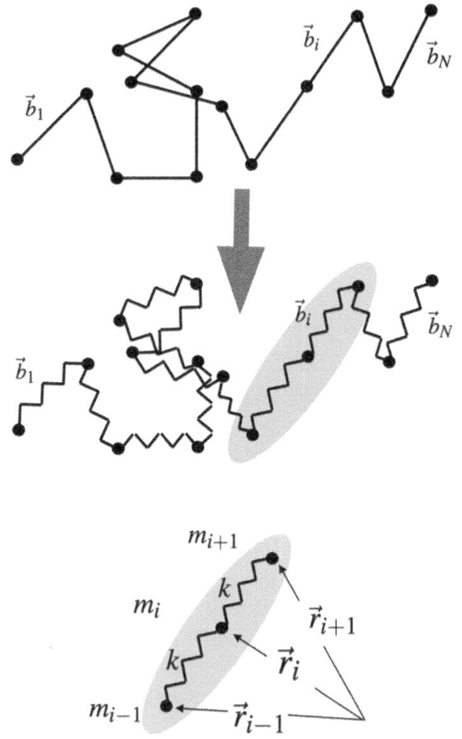

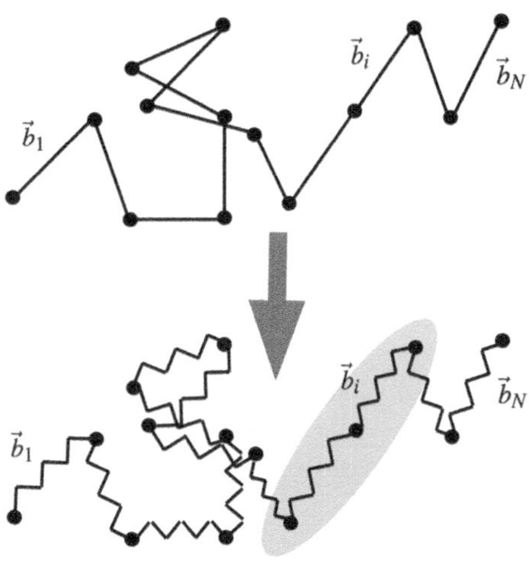

Fig. 4.19 Illustration of the Rouse model

Next we want to express the $\vec{r}(i, t)$ in terms of new coordinates $\hat{\vec{r}}(p, t)$ so that in these coordinates (4.105) assumes the form

$$\frac{d}{dt}\hat{\vec{r}}(p, t) = -\frac{k_p}{\zeta_p}\hat{\vec{r}}(p, t) + \frac{1}{\zeta_p}\hat{\vec{Z}}(p, t). \tag{4.108}$$

This equation is the three-dimensional version of (4.98), which we found when we studied the dynamics of a one-dimension oscillator. If we succeed, it means that we have found normal mode coordinates for the Rouse chain, i.e. we can study its dynamics in terms of independent harmonic oscillators for each mode.

Following Doi and Edwards we consider the linear transformation

$$\hat{\vec{r}}(p, t) = \int_0^N di \, \phi(p, i) \vec{r}(i, t). \tag{4.109}$$

Hence,

$$\frac{d}{dt}\hat{\vec{r}}(p, t) = \int_0^N di \, \phi(p, i) \left(\frac{k}{\zeta}\frac{d^2\vec{r}(i, t)}{di^2} + \frac{1}{\zeta}\vec{Z}(i, t) \right). \tag{4.110}$$

Using repeated partial integration we can manipulate the term containing the double derivative d^2/di^2, which yields

$$\int_0^N di\,\phi(p,i)\frac{d^2\vec{r}(i,t)}{di^2} = \int_0^N di\,\frac{d^2\phi(p,i)}{di^2}\vec{r}(i,t) - \Big|_0^N \frac{d\phi(p,i)}{di}\vec{r}(i,t). \qquad (4.111)$$

Here we have used the boundary condition (4.107) upon the first partial integration. Substituting (4.111) into (4.110) we see that we indeed obtain (4.108) if the following conditions are satisfied:

$$\frac{d^2}{di^2}\phi(p,i) = -\frac{\zeta/k}{\zeta_p/k_p}\phi(p,i) \qquad (4.112)$$

$$\frac{d}{di}\phi(p,i) = 0 \qquad (i=0,N) \qquad (4.113)$$

$$\hat{\vec{Z}}(p,t) = \frac{\zeta_p}{\zeta}\int_0^N di\,\phi(p,i)\,\vec{Z}(i,t). \qquad (4.114)$$

A solution satisfying both (4.112) and (4.113) is

$$\phi(p,i) = \frac{1}{N}\cos\left(\frac{p\pi i}{N}\right) \quad,\text{ where } \quad \frac{p^2\pi^2}{N^2} = \frac{\zeta/k}{\zeta_p/k_p}. \qquad (4.115)$$

Finally[5] we shall need $\langle \hat{\vec{Z}}(p,t') \cdot \hat{\vec{Z}}(q,t'')\rangle$ when we study the dynamics of the Rouse chain. Using (4.114) we have

$$\langle \hat{\vec{Z}}(p,t') \cdot \hat{\vec{Z}}(q,t'')\rangle = \frac{\zeta_p\zeta_q}{\zeta^2 N^2}\int_0^N di \int_0^N dj\,\langle \vec{Z}(i,t') \cdot \vec{Z}(j,t'')\rangle \qquad (4.118)$$
$$\times \cos\left(\frac{p\pi i}{N}\right)\cos\left(\frac{q\pi j}{N}\right).$$

Inserting

$$\langle \vec{Z}(i,t') \cdot \vec{Z}(j,t'')\rangle = 6\zeta k_B T\delta(t'-t'')\delta_{i,j} \qquad (4.119)$$

---

[5] The inverse transform of (4.109) with $\phi(p,i)$ given by (4.115) is

$$\vec{r}(i,t) = \hat{\vec{r}}(0,t) + 2\sum_{q=1}^N \hat{\vec{r}}(q,t)\cos\left(\frac{q\pi i}{N}\right). \qquad (4.116)$$

We can check this by inserting $\hat{\vec{r}}(q,t)$ given by (4.109) into this equation and using

$$\int_0^N di\,\cos\left(\frac{q\pi i}{N}\right)\cos\left(\frac{p\pi i}{N}\right) = \frac{N}{2}\delta_{q,p}(1+\delta_{q,0}). \qquad (4.117)$$

(cf. (4.95)) the integration yields

$$\langle \hat{\dot{Z}}(p,t') \cdot \hat{\dot{Z}}(q,t'') \rangle = 6\frac{\zeta_p^2}{\zeta N^2} N k_B T \delta(t' - t'')\delta_{p,q}\left(\frac{1}{2} + \frac{1}{2}\delta_{0,q}\right). \tag{4.120}$$

We do have some freedom of choice in the case of ζ_p and use it to require that the right hand sides in (4.119) and (4.120) have the same form (with ζ in (4.119) becoming ζ_p in (4.120)). This in turn implies

$$\zeta_0 = \zeta N \quad \text{and} \quad \zeta_p = 2\zeta N \quad (p > 0), \tag{4.121}$$

i.e.

$$k_p = \frac{6\pi^2 k_B T}{b^2 N} p^2 \quad (p > 0). \tag{4.122}$$

Finally, we adopt the two results (4.102) and (4.103) from the one-dimensional oscillator example to the Rouse chain, i.e.

$$\langle \hat{r}_\alpha(p,0)\hat{r}_\beta(q,t) \rangle = \delta_{p,q}\delta_{\alpha,\beta}\frac{k_B T}{k_p} \exp[-t/\tau_p'] \quad (p = 1, 2, \ldots), \tag{4.123}$$

where

$$\tau_p' \equiv \frac{\zeta_p}{k_p} \tag{4.124}$$

and

$$\langle (\hat{r}_\alpha(0,t) - \hat{r}_\alpha(0,0))(\hat{r}_\beta(0,t) - \hat{r}_\beta(0,0)) \rangle = 2\delta_{\alpha,\beta}\frac{k_B T}{\zeta_0}t. \tag{4.125}$$

Now we can discuss predictions of the Rouse model:

- *Center of mass diffusion of the Rouse chain:*

Equation (4.109) immediately tells us that $\hat{\vec{r}}(0,t)$ is the center of mass of the chain. Hence, (4.125) tells us what the mean square displacement of the center of mass is and, in particular, its diffusion coefficient:

$$D_{cm} = \frac{k_B T}{\zeta_0} = \frac{k_B T}{\zeta N}. \tag{4.126}$$

● *Rotational relaxation of the Rouse chain:*

We use the relaxation time τ_r of the auto-correlation function of the end-to-end vector, $\langle \vec{R}(0) \cdot \vec{R}(t) \rangle \sim \exp[-t/\tau_r]$, as a measure for rotational relaxation of the chain. Since

$$\vec{R}(t) = \vec{r}(N, t) - \vec{r}(0, t) \overset{(4.116)}{=} -4 \sum_{p=1 \,(p \text{ odd})}^{N} \hat{\vec{r}}(p, t), \tag{4.127}$$

we find via (4.123)

$$\langle \vec{R}(0) \cdot \vec{R}(t) \rangle = 16 \sum_{p=1 \,(p \text{ odd})}^{N} \frac{3k_B T}{k_p} \exp[-t/\tau'_p]. \tag{4.128}$$

Obviously this is a sum over many exponentials. But of all the relaxation times

$$\tau'_p = \frac{\zeta_p}{k_p} = \frac{\zeta b^2 N^2}{3\pi^2 p^2 k_B T} \tag{4.129}$$

τ'_1 is the longest. This one we identify with τ_r:

$$\tau_r \approx \frac{\zeta_1}{k_1} = \frac{\zeta b^2 N^2}{3\pi^2 k_B T}. \tag{4.130}$$

● *Storage and loss modulus in the Rouse model:*

In (4.75) we had introduced the shear relaxation modulus $G(t)$ from which we can calculate $G'(\omega)$ and $G''(\omega)$ using (4.83) and (4.84). But how do we find $G(t)$? A formal derivation starting from the microscopic stress tensor can be found in Doi and Edwards (Chap. 4). Here we only want to hazard a guess at a plausible form of $G(t)$.

Note that the unit of G is energy per volume. In the limit $t \to 0$ all modes contribute, since none has relaxed yet. The relevant energy is $k_B T \sum_p$ times the number of chains inside the volume, i.e.

$$G(0) = \rho k_B T \sum_p 1. \tag{4.131}$$

The quantity ρ is the number of chains per volume. If we now consider the time behavior of $G(t)$, we may be tempted to simply assume that it follows the same auto-correlation function as the normal mode coordinates in (4.123). However, as was pointed out above, the unit of stress is energy per volume and, when we think back to the one-dimensional oscillator example, the energy is build from squares of the coordinates. Hence, rather than using $\langle x(0)x(t)\rangle$, referring to the one-dimensional oscillator, we need to base our guess on $\langle x^2(0)x^2(t)\rangle$.[6] The resulting conclusion is

$$G(t) = \rho k_B T \sum_p \exp[-2t/\tau_p']. \tag{4.134}$$

Inserting this into (4.83) and (4.84) and integrating term by term yields

$$G'(\omega) = \rho k_B T \sum_p \frac{(\omega \tau_p'')^2}{(\omega \tau_p'')^2 + 1} \tag{4.135}$$

and

$$G''(\omega) = \rho k_B T \sum_p \frac{\omega \tau_p''}{(\omega \tau_p'')^2 + 1} \tag{4.136}$$

$(\tau_p'' = \tau_p'/2)$.

Let us consider this result in its two limits, i.e. $\omega\tau_1'' \ll 1$ and $\omega\tau_1'' \gg 1$ (Note that we use the longest relaxation time!). In the case $\omega\tau_1'' \ll 1$

$$\sum_p \frac{(\omega \tau_p'')^2}{(\omega \tau_p'')^2 + 1} \approx \omega^2 \tau_1''^2 \underbrace{\sum_{p=1}^{\infty} p^{-4}}_{=\pi^4/90\approx 1.1} \tag{4.137}$$

and

$$\sum_p \frac{(\omega \tau_p'')}{(\omega \tau_p'')^2 + 1} \approx \omega \tau_1'' \underbrace{\sum_{p=1}^{\infty} p^{-2}}_{=\pi^2/6\approx 1.6}. \tag{4.138}$$

---

[6] The calculation of $\langle x^2(0)x^2(t)\rangle$ is analogous to that of $\langle x(0)x(t)\rangle$ (cf. (4.102)). The main exception is that we need to deal with the 4-point function $\langle Z_x(t_1)Z_x(t_2)Z_x(t_3)Z_x(t_4)\rangle$, which can be expressed as a sum over products of 2-point functions:

$$\langle Z_x(t_1)Z_x(t_2)Z_x(t_3)Z_x(t_4)\rangle = \langle Z_x(t_1)Z_x(t_2)\rangle\langle Z_x(t_3)Z_x(t_4)\rangle \tag{4.132}$$
$$+\langle Z_x(t_1)Z_x(t_3)\rangle\langle Z_x(t_2)Z_x(t_4)\rangle + \langle Z_x(t_1)Z_x(t_4)\rangle\langle Z_x(t_2)Z_x(t_3)\rangle.$$

(cf. [18]; p. 192). With this we obtain

$$\langle x^2(0)x^2(t)\rangle - \langle x^2\rangle^2 = \frac{2k_B^2 T^2}{k^2} \exp[-2t/\tau']. \tag{4.133}$$

Note that $\langle x^2\rangle^2$ (aside from a factor) is the square of the thermal energy of the oscillator which we subtract.

Table 4.3 Limiting ω-dependence of $G'(\omega)$ and $G''(\omega)$ predicted by the Rouse model

	$\omega\tau_1'' \ll 1$	$\omega\tau_1'' \gg 1$
$G'(\omega)$	$\sim\omega^2$	$\sim\omega^{1/2}$
$G''(\omega)$	$\sim\omega$	$\sim\omega^{1/2}$

In the opposite limit, i.e. $\omega\tau_1'' \gg 1$, we can replace the summation by an integration. Hence

$$\sum_p \frac{(\omega\,\tau_p'')^2}{(\omega\,\tau_p'')^2 + 1} \approx \int_0^\infty dp \frac{(\omega\,\tau_1'')^2}{(\omega\,\tau_1'')^2 + p^4} = \frac{\pi}{2\sqrt{2}}(\omega\,\tau_1'')^{1/2} \qquad (4.139)$$

and

$$\sum_p \frac{(\omega\,\tau_p'')}{(\omega\,\tau_p'')^2 + 1} \approx \int_0^\infty dp \frac{\omega\,\tau_1''\,p^2}{(\omega\,\tau_1'')^2 + p^4} = \frac{\pi}{2\sqrt{2}}(\omega\,\tau_1'')^{1/2}. \qquad (4.140)$$

The limiting ω-dependencies of $G'(\omega)$ and $G''(\omega)$ are compiled in Table 4.3.

Finally, note that every term in the sums in (4.135) and (4.136) corresponds to a Maxwell model, one of the phenomenological models discussed in Appendix A— in principle a justification, albeit with limitations, for using this phenomenological model.

Figures 4.20 and 4.21 show examples comparing experimental data for the dynamic moduli to theoretical predictions. The experimental data points in Fig. 4.20 are taken from Fig. 4 in [19]. The authors present their results for the dynamic viscoelasticity of dilute polyelectrolyte solutions, where the polymer is poly(2-vinylpyridine). The solid lines are $G'/(\rho k_B T)$ and $G''/(\rho k_B T)$ from (4.135) and (4.136) using $\tau_p'' = K/p^2$ and $K = 0.7$, i.e. K is the only adjustable parameter. The overall agreement between the experiment and the Rouse prediction is quite reasonable. However, there are many points here which we do not address. For a detailed discussion the reader is referred to the original publication. The additional dashed lines correspond to Zimm's modification of the Rouse model, which we discuss in the following. The experimental data points in Fig. 4.21 are taken from Fig. 5 in [20] for the case of two different good solvents ($\nu = 3/5$). As in the previous comparison, the solid lines are $gG'/(\rho k_B T)$ and $gG''/(\rho k_B T)$ from (4.135) and (4.136) $\tau_p'' = K/p^2$, and $K = 0.008$. Here an additional factor $g = 110$ was used to shift the curves. The dashed lines again correspond to Zimm's modification of the Rouse model.

The Rouse model describes G' and G'' of polymers in dilute solution and, as we shall see, of 'short' polymers in the melt reasonably well ('Reasonably well' may not always apply. Nevertheless, in a rough sense the statement is a fair assessment.). However, both the center of mass diffusion as well as the rotational diffusion, specifically their dependence on N, are generally incorrect. But why? Even though the beads in the Rouse chain are mechanically coupled, every bead interacts indepen-

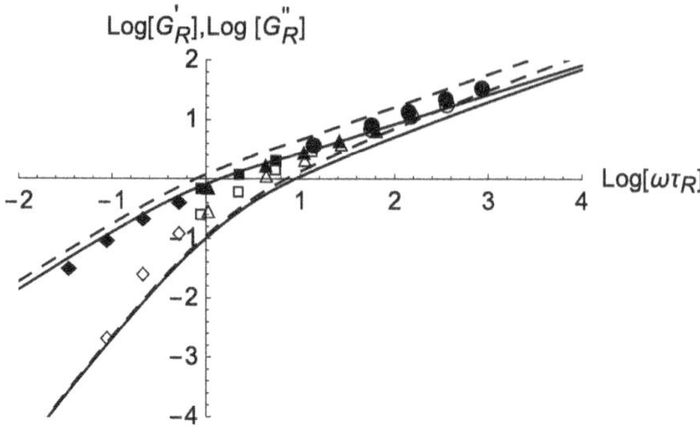

Fig. 4.20 Reduced storage modulus (open symbols) and reduced loss modulus (closed symbols) versus ω times a reduced relaxation time. Data from Fig. 4 in [19]. The different symbols indicate molecular weights ranging from $3 \cdot 10^4$ to $1 \cdot 10^6$ g/mol in 0.0023 M HCl/ethylene glycol at a concentration of 2.0 mg mL^{-1}. The solid lines are $G'/(\rho k_B T)$ and $G''/(\rho k_B T)$ from (4.135) and (4.136) using $\tau_p'' = K/p^2$ and $K = 0.7$. Dashed lines are $G'/(\rho k_B T)$ and $G''/(\rho k_B T)$ obtained with the Zimm model. In this case $\tau_p'' = K/p^{3\nu}$ ($K = 0.7$ and $\nu = 3/5$)

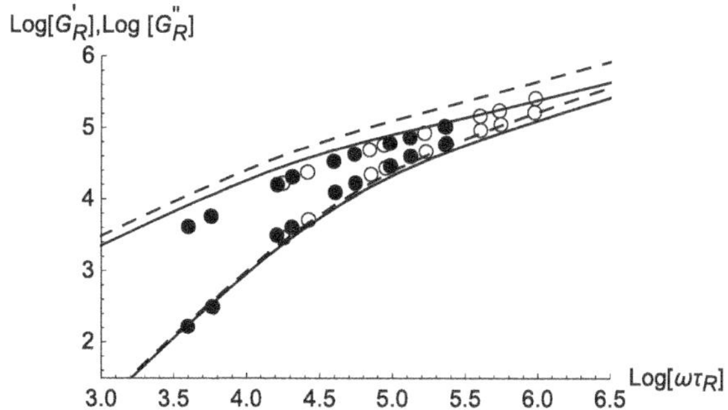

Fig. 4.21 Reduced storage modulus (bottom symbols) and reduced loss modulus (top symbols) versus ω times a reduced relaxation time. Data from Fig. 5 in [20]. The solid lines are $gG'/(\rho k_B T)$ and $gG''/(\rho k_B T)$ from (4.135) and (4.136) using $g = 110$, $\tau_p'' = K/p^2$, and $K = 0.008$. Dashed lines are $G'/(\rho k_B T)$ and $G''/(\rho k_B T)$ obtained with the Zimm model. In this case $\tau_p'' = K/p^{3\nu}$ ($K = 0.008$ and $\nu = 3/5$)

dently with the surrounding solvent. The drag force on the Rouse chain is therefore the sum over the drag forces on individual beads. Consequently the overall drag coefficient scales according to N. The Rouse model neglects the hydrodynamical coupling of the beads. In the following we shall fix this problem and we shall see that the main difference is that the chain in a solvent diffuses and rotates as one 'object' possessing linear dimension proportional to N^ν.

We began our discussion of chain dynamics by looking at a sphere moving within a liquid while experiencing Stoke's friction proportional to its velocity. Each bead in the Rouse model is such a sphere, which up to now we have treated as hydrodynamically independent. This is not correct. Instead we express the force a bead i exerts on its surroundings due to **hydrodynamic interaction** as

$$\vec{F}_i = 6\pi\eta R\left[\vec{v}_i - \vec{v}(\vec{r}_i)\right] \equiv \zeta\left[\vec{v}_i - \vec{v}(\vec{r}_i)\right]. \tag{4.141}$$

Here $\vec{v}_i$ is the velocity of bead i in the 'laboratory' frame and $\vec{v}(\vec{r})$ is the flow field in the same frame created by the sum of the corresponding forces $\vec{\varphi}(\vec{r})$ of all other beads, i.e.

$$\vec{\varphi}(\vec{r}) = \sum_{j(\neq i)} \delta(\vec{r} - \vec{r}_j)\vec{F}_j(\vec{r}_j). \tag{4.142}$$

The equation linking $\vec{v}_i$ to $\vec{\varphi}(\vec{r})$ is (4.45). Hence,

$$\vec{v}_i = \frac{\vec{F}_i}{\zeta} + \vec{v}(\vec{r}_i) \tag{4.143}$$

$$= \frac{\vec{F}_i}{\zeta} + \int d^3r'\,\mathbf{H}(\vec{r}_i - \vec{r}\,')\vec{\varphi}(\vec{r}\,')$$

$$= \frac{\vec{F}_i(\vec{r}_i)}{\zeta} + \sum_{j(\neq i)} \mathbf{H}(\vec{r}_i - \vec{r}_j)\vec{F}_j(\vec{r}_j)$$

or

$$\vec{v}_i = \sum_j \mathbf{T}(i, j)\vec{F}_j, \tag{4.144}$$

where

$$\mathbf{T}(i, j) = \begin{cases} \frac{\mathbf{I}}{\zeta} & (i = j) \\ \mathbf{H}(\vec{r}_i - \vec{r}_j) & (i \neq j) \end{cases} \tag{4.145}$$

is the **mobility matrix** ($\mathbf{I}$ is the unit matrix.).

Applying (4.144) to the Rouse equation (4.105), we see that in order to include hydrodynamic interactions we must modify (4.105) to

$$\frac{d}{dt}\vec{r}(i, t) = \sum_j \mathbf{T}(i, j)\left(k\frac{d^2\vec{r}(j, t)}{dj^2} + \vec{Z}(j, t)\right), \tag{4.146}$$

where

$$\langle Z_{i,\alpha} Z_{j,\beta}\rangle = 2(\mathbf{T})^{-1}_{\alpha\beta}(i, j)k_B T\,\delta(t - t'). \tag{4.147}$$

Note that $\mathbf{T}(i, j)$ does depend on $\vec{r}_j$, which means that (4.146) is a nonlinear equation. Zimm avoided this difficulty by replacing $\mathbf{T}(i, j)$ with its equilibrium average $\langle \mathbf{T}(i, j) \rangle$, i.e.

$$\langle \mathbf{T}(i, j) \rangle = \int d^3 r_{ij}\, p(r_{ij}) \mathbf{T}(i, j), \tag{4.148}$$

where $p(r_{ij})$ is, for instance, the ideal chain distribution (2.67). This step, called **pre-averaging** in the literature, yields the **Zimm model** [16], i.e.

$$\frac{d}{dt}\vec{r}(i, t) = \int_0^N dj\, \langle \mathbf{T}(i, j) \rangle \left(k \frac{d^2 \vec{r}(j, t)}{dj^2} + \vec{Z}(j, t) \right), \tag{4.149}$$

where additionally the sum over j is replaced by an integration.

Instead of (2.67), the formula for $p(r_{ij})$ for the ideal chain, we are going to use (2.116) to evaluate $\langle \mathbf{T}(i, j) \rangle$. Remember that we found (2.116) by using scaling ideas and that this distribution function applies to ideal as well as real chains. Hence,

$$\langle T_{\alpha\beta}(i, j) \rangle = \frac{\int d^3 r\, \exp[-k_\nu \left(\frac{r}{b|i-j|^\nu} \right)^{\frac{1}{1-\nu}}] \frac{\delta_{\alpha\beta}+e_{r,\alpha}e_{r,\beta}}{8\pi\eta r}}{\int d^3 r\, \exp[-k_\nu \left(\frac{r}{b|i-j|^\nu} \right)^{\frac{1}{1-\nu}}]}. \tag{4.150}$$

Note[7] that

$$\langle e_{r,\alpha}e_{r,\beta} \rangle = \frac{1}{3}\delta_{\alpha\beta}. \tag{4.152}$$

The r-integrations can be done with the help of, for instance, Mathematica. The final result is

$$\langle \mathbf{T}(i, j) \rangle = t(i - j)\, \mathbf{I} \tag{4.153}$$

with

$$t(i - j) = \frac{1}{6\pi\eta} \frac{\Gamma[2(1 - \nu)]}{\Gamma[3(1 - \nu)]} \frac{k_\nu^{1-\nu}}{b|i - j|^\nu}. \tag{4.154}$$

In the case of an ideal chain ($\nu = 1/2$ and $k_\nu = 3/2$) $t(i - j)$ becomes

---

[7] k_ν can be determined via $\langle r^2 \rangle = b^2 N^{2\nu}$, which yields

$$k_\nu = \left(\frac{\Gamma[5(1 - \nu)]}{\Gamma[3(1 - \nu)]} \right)^{\frac{1}{2(1-\nu)}} = \begin{cases} 3/2 & (\nu = 1/2) \\ \approx 1.11 & (\nu = 3/5) \end{cases}. \tag{4.151}$$

Here $\Gamma[x]$ is the Gamma function (e.g., Chap. 6 in [17]; see also Problem 3.1).

$$t(i-j) \stackrel{id.ch.}{=} \frac{1}{\eta(6\pi^3 b^2 |i-j|)^{1/2}}. \tag{4.155}$$

The equivalent to the Rouse equation in normal mode coordinates, i.e. (4.108), is here

$$\frac{d}{dt}\hat{\vec{r}}(p,t) = \sum_q \hat{t}_{pq}\left(-k_q\hat{\vec{r}}(q,t) + \hat{\vec{Z}}(q,t)\right), \tag{4.156}$$

where k_q is given in (4.115) and

$$\hat{t}_{pq} = \frac{1}{N^2}\int_0^N di \int_0^N dj \cos\left(\frac{p\pi i}{N}\right)\cos\left(\frac{q\pi j}{N}\right)t(i-j). \tag{4.157}$$

In the case $p=0$ or $q=0$ the result is

$$\hat{t}_{0q} = \hat{t}\,\delta_{0,q} \quad \text{or} \quad \hat{t}_{p0} = \hat{t}\,\delta_{p,0} \tag{4.158}$$

with

$$\hat{t} = \frac{1}{3\pi\eta}\frac{k_\nu^{1-\nu}\Gamma[2(1-\nu)]}{(1-\nu)(2-\nu)\Gamma[3(1-\nu)]}\frac{1}{bN^\nu}. \tag{4.159}$$

In the case $p,q>0$ we resort to an approximate solution. Using the substitution $j-i=l$ we have

$$\frac{1}{N^2}\int_0^N di \int_{-i}^{N-i} dl \cos\left(\frac{p\pi i}{N}\right)\cos\left(\frac{q\pi(l+i)}{N}\right)|l|^{-\nu}$$

$$\approx \frac{1}{N^2}\underbrace{\int_0^N di \cos\left(\frac{p\pi i}{N}\right)\cos\left(\frac{q\pi i}{N}\right)}_{=(N/2)\delta_{p,q}}\underbrace{\int_{-\infty}^{\infty} dl \cos\left(\frac{q\pi l}{N}\right)|l|^{-\nu}}_{=2(N/(\pi q))^{1-\nu}\Gamma[1-\nu]\sin(\pi\nu/2)}$$

$$-\frac{1}{N^2}\int_0^N di \cos\left(\frac{p\pi i}{N}\right)\sin\left(\frac{q\pi i}{N}\right)\underbrace{\int_{-\infty}^{\infty} dl \sin\left(\frac{q\pi l}{N}\right)|l|^{-\nu}}_{=0}.$$

Hence, for $p,q>0$

$$\hat{t}_{pq} \approx \frac{1}{6\pi\eta}\frac{\Gamma[1-\nu]\Gamma[2(1-\nu)]}{\Gamma[3(1-\nu)]}\left(\frac{k_\nu}{\pi p}\right)^{1-\nu}\sin\left(\frac{\pi\nu}{2}\right)\frac{1}{bN^\nu}\delta_{p,q}. \tag{4.160}$$

Note that for $\nu=1/2$

$$\hat{t}_{00} = \frac{8}{3(6\pi^3)^{1/2}\eta b\sqrt{N}} \tag{4.161}$$

and

$$\hat{t}_{11} \approx \frac{1}{\sqrt{2}(6\pi^3)^{1/2}\eta b\sqrt{N}}.$$
(4.162)

We observe that due to the δ_{pq} in the formulas (4.158) and (4.160), (4.156) truly assumes the form of its counterpart, (4.108), in the Rouse model, i.e.

$$\frac{d}{dt}\hat{\vec{r}}(p,t) = \hat{t}_{pp}\left(-k_p\hat{\vec{r}}(p,t) + \hat{\vec{Z}}(p,t)\right).$$
(4.163)

Hence we directly obtain Zimm's version of our previous Rouse results:

- *Center of mass diffusion of the Zimm chain:*

Since $1/\zeta_o = \hat{t}_{00}$, we find for the case $\nu = 1/2$ using (4.161)

$$D_{cm} = k_B T \hat{t}_{00} = \frac{8k_B T}{3(6\pi^3)^{1/2}\eta b\sqrt{N}}.$$
(4.164)

In general, i.e. for general ν, we see from (4.159) that

$$D_{cm} \propto N^{-\nu}.$$
(4.165)

This makes good physical sense. Our starting point was Stoke's friction with a drag coefficient ζ proportional to the radius R of the sphere. The linear dimension of a polymer chain is proportional to N^ν and thus the corresponding drag coefficient is expected to be proportional to N^ν. This type of reasoning of course depends on the range of the hydrodynamic interaction, which must be greater than the polymer dimension. However, the r^{-1}-dependence in the Oseen tensor means that the hydrodynamic interaction is virtually infinite.

- *Rotational relaxation of the Zimm chain:*

We seek the rotation relaxation time τ_r in the case of the Zimm model. According to (4.130)

$$\tau_r \approx \tau_1' = \frac{\zeta_1}{k_1}.$$
(4.166)

Here $\zeta_1 = 1/\hat{t}_{11}$. $\hat{t}_{11}$ is given in (4.162) for the ideal chain and in the general case we can use (4.160). But what about k_1? Does (4.122), derived in the context of the Rouse model, still apply when $\nu = 1/2$ and what is k_1 when the chain is a real chain, i.e. $\nu = 3/5$? One answer is that (4.122) still applies when $\nu = 1/2$. In order to verify this statement and, in addition, find the answer to the second question we need to calculate k_p explicitly via the formula

$$\langle \hat{\vec{r}}(p)^2 \rangle = \frac{3k_B T}{k_p}. \tag{4.167}$$

Note that k_p is the force constant of an individual uncoupled oscillator. The above formula gives the position fluctuations of this oscillator (mode) for a particular temperature. For example, let us consider the one-dimensional oscillator with the potential energy (4.97). In this case

$$\langle x^2 \rangle = \frac{\int_{-\infty}^{\infty} dx\, x^2 \exp[-\beta k x^2/2]}{\int_{-\infty}^{\infty} dx \exp[-\beta k x^2/2]} = \frac{1}{\beta k}, \tag{4.168}$$

where $\beta = (k_B T)^{-1}$. The factor 3 in (4.167) results since the oscillator is three-dimensional.

For a reason not immediately obvious we do not use

$$\hat{\vec{r}}(p,t) = \frac{1}{N} \int_0^N di \cos\left(\frac{p\pi i}{N}\right) \vec{r}(i,t) \tag{4.169}$$

directly but

$$\hat{\vec{r}}(p,t) = -\frac{1}{p\pi} \int_0^N di \sin\left(\frac{p\pi i}{N}\right) \frac{d\vec{r}(i,t)}{di} \tag{4.170}$$

instead, which follows from (4.169) via partial integration. Hence,

$$\langle \hat{\vec{r}}(p)^2 \rangle = \frac{1}{p^2\pi^2} \int_0^N di \int_0^N dj \sin\left(\frac{p\pi i}{N}\right)$$
$$\times \sin\left(\frac{p\pi j}{N}\right) \langle \frac{d\vec{r}(i,t)}{di} \cdot \frac{d\vec{r}(j,t)}{dj} \rangle. \tag{4.171}$$

Now we use

$$\frac{d\vec{r}(i,t)}{di} \cdot \frac{d\vec{r}(j,t)}{dj} = -\frac{1}{2}\frac{d^2}{di\,dj}(\vec{r}(i,t) - \vec{r}(j,t))^2 \tag{4.172}$$

in conjunction with

$$\langle (\vec{r}(i,t) - \vec{r}(j,t))^2 \rangle = b^2 |i - j|^{2\nu}, \tag{4.173}$$

which yields

$$\langle \frac{d\vec{r}(i,t)}{di} \cdot \frac{d\vec{r}(j,t)}{dj} \rangle = (2\nu - 1)\nu b^2 |i - j|^{2(\nu-1)}. \tag{4.174}$$

Substituting this back into (4.171), we are left with an integration just like the one in the case of $\hat{t}_{pq}$. Using as before the substitution $j - i = l$ and the subsequent extension of the l-integration limits to $\pm\infty$ we obtain (approximately)

$$\langle \hat{\vec{r}}(p)^2 \rangle = \frac{\nu \Gamma[2\nu]}{\pi^{2\nu+1}} \sin(\pi(1 - \nu)) \frac{b^2 N^{2\nu}}{p^{2\nu+1}} (1 - \delta_{0p}). \tag{4.175}$$

Inserting $\nu = 1/2$ into this formula and using (4.167) yields, as previously formulated as a question, (4.122) for k_p. More generally we obtain,

$$k_p \sim \frac{k_B T p^{2\nu+1}}{b^2 N^{2\nu}}. \tag{4.176}$$

Hence, for $\nu = 1/2$

$$\tau_r = \tau_1' = \frac{\zeta_1}{k_p} = \frac{\eta(bN^{1/2})^3}{\sqrt{3\pi}k_B T} \approx 0.33 \frac{\eta(bN^{1/2})^3}{k_B T} \tag{4.177}$$

and for $\nu = 3/5$

$$\tau_r = \tau_1' \approx 0.18 \frac{\eta(bN^{3/5})^3}{k_B T}, \tag{4.178}$$

and generally

$$\tau_p' \sim \frac{\eta(bN^\nu/p^\nu)^3}{k_B T}. \tag{4.179}$$

If we compare this result of the Zimm model with τ_p' in the Rouse model (4.129), we find

$$\frac{\tau_p'(\text{Rouse})}{\tau_p'(\text{Zimm})} \sim \left(\frac{N}{p}\right)^{2-3\nu}, \tag{4.180}$$

where $2 - 3\nu$ is equal to $1/2$ $(1/5)$ for $\nu = 1/2$ $(3/5)$. This means that long wavelength modes in particular (small p) have significantly longer relaxation times in the Rouse model compared to the Zimm model.

• *Storage and loss modulus in the Zimm model:*

The functional form of the moduli in the (4.137) and (4.138) remains the same. What is different is τ'_p and therefore $\tau''_p = \tau'_p/2$. This means that the two limits, i.e. $\omega\tau''_1 \ll 1$ and $\omega\tau''_1 \gg 1$, and thus (4.137) through (4.140) require modification:

In the case $\omega\tau''_1 \ll 1$

$$\sum_p \frac{(\omega\tau''_p)^2}{(\omega\tau''_p)^2 + 1} \approx \omega^2\tau''^2_1 \underbrace{\sum_{p=1}^{\infty} p^{-(6\nu)}}_{=\zeta(6\nu)=\begin{cases} \approx 1.2\ \nu = 1/2 \\ \approx 1.1\ \nu = 3/5 \end{cases}} \tag{4.181}$$

and

$$\sum_p \frac{(\omega\tau''_p)}{(\omega\tau''_p)^2 + 1} \approx \omega\tau''_1 \underbrace{\sum_{p=1}^{\infty} p^{-(3\nu)}}_{=\zeta(3\nu)=\begin{cases} \approx 2.6\ \nu = 1/2 \\ \approx 1.9\ \nu = 3/5 \end{cases}} \cdot \tag{4.182}$$

In the opposite limit, i.e. $\omega\tau''_1 \gg 1$,

$$\sum_p \frac{(\omega\tau''_p)^2}{(\omega\tau''_p)^2 + 1} \approx \int_0^{\infty} dp\, \frac{(\omega\tau''_1)^2}{(\omega\tau''_1)^2 + p^{6\nu}} = \tag{4.183}$$

$$\underbrace{\frac{\pi}{6\nu}\csc(\frac{\pi}{6\nu})}\ (\omega\tau''_1)^{1/(3\nu)}$$

$$= \begin{cases} \approx 1.2\ \nu = 1/2 \\ \approx 1.1\ \nu = 3/5 \end{cases}$$

and

$$\sum_p \frac{\omega\tau''_p}{(\omega\tau''_p)^2 + 1} \approx \int_0^{\infty} dp\, \frac{\omega\tau''_1 p^{3\nu}}{(\omega\tau''_1)^2 + p^{6\nu}} = \tag{4.184}$$

$$\underbrace{\frac{\pi}{6\nu}\sec(\frac{\pi}{6\nu})}\ (\omega\tau''_1)^{1/(3\nu)}.$$

$$= \begin{cases} \approx 2.1\ \nu = 1/2 \\ \approx 1.4\ \nu = 3/5 \end{cases}$$

Table 4.4 Limiting ω-dependence of $G'(\omega)$ and $G''(\omega)$ predicted by the Zimm model

	$\omega\tau_1'' \ll 1$	$\omega\tau_1'' \gg 1$
$G'(\omega)$	$\sim\omega^2$	$\sim\omega^{1/(3\nu)}$
$G''(\omega)$	$\sim\omega$	$\sim\omega^{1/(3\nu)}$

The ω dependencies of $G'(\omega)$ and $G''(\omega)$ in the Zimm model are compiled in Table 4.4.

Aside from small differences in the factors, the ω-dependence in the low frequency limit is the same as in the Rouse case. In the high frequency limit the power of $\omega\tau_1''$ is different and in the Zimm model depends on ν. However, as the comparisons in the Figs. 4.20 and 4.21 show, the dynamic moduli computed with the Zimm model for $\nu = 3/5$ are not very different from the corresponding Rouse results. Note however that no particular effort is made to 'tweak' the parameter K, which is different in the two models but here is assumed to be the same to keep things simple. Note also in this context that the dependence of τ_1'' on N in the two models is different, i.e.

$$\tau_1'' \sim \begin{cases} N^2 & \text{Rouse} \\ N^{3\nu} & \text{Zimm} \end{cases} . \tag{4.185}$$

4.3 Entanglement

Figure 4.22 is a copy of Fig. 4.13 in which the red lines possess slopes predicted by the Zimm model for the storage modulus G' at low and high frequencies, respectively. Note that the experimental system is a melt and thus $\nu = 1/2$. The frequency dependence of G' observed previously for polymers in dilute solution, i.e. in Figs. 4.20 and 4.21, here is seen only when the chains are short. The limiting molecular masses in Fig. 4.22 correspond to about $0.9 \cdot 10^2$ and roughly $6 \cdot 10^3$ monomers per polymer chain, respectively. It appears that when the chains exceed a certain length, somewhere above 10^2 monomers, the two limits become separated by a third frequency range in which the storage modulus is constant and independent of molecular weight (or chain length). Can we explain the occurrence of this so-called **plateau modulus**?

From the point of view of the individual chain its environment is an 'entangled mess' of other chains. How does this chain move? Figure 4.23 is a cartoon of an idea which has proven useful with regard to this question. The central part of the cartoon depicts a polymer chain consisting of N monomers (or perhaps Kuhn segments). Again there are statistical 'blobs' containing on average N_e monomers along this chain stuck inside a **'tube'**. The diameter of the tube corresponds to the linear dimension of the blobs. But what defines the length N_e and thus the linear dimension of the blobs? It is assumed that this dimension is determined by 'entanglements', i.e.

interlaced 'hairpins' formed by the polymer with its neighbors (or even with itself). Hence the polymer contour length between hairpins is N_e, the **entanglement length**. When the polymer moves, 'trying' to free itself from the entanglements, its motion is akin to a snake moving forward via a looping motion of its body. Hence the term **'reptation'** was coined in this context by de Gennes. Note that the lower left corner in Fig. 4.23 is a closeup view of a section of the aforementioned tube. The entire tube contains and is made up by the entire polymer chain. This tube itself is a random walk with a characteristic end-to-end distance or linear dimension R_{tube}. In the following we exploit this picture using the previously developed scaling concept.

The entanglements subdivide the entire polymer chain into N/N_e 'pieces'. To the N_e effective monomers or Kuhn segments between two entanglements we apply the formulas (2.78) and (2.87) (polymer inside a tube), where $N_\xi = N_e$ and $\xi = D_e$. Here D_e, given by

$$D_e \sim N_e^\nu, \tag{4.186}$$

is the diameter of a tube along which the polymer chain reptates. The tube's contour length, i.e. the length of tube containing the polymer, is

$$L = \frac{N}{N_e} D_e. \tag{4.187}$$

The time τ_{rep} it takes for the polymer to reptate this length we describe in terms of Einsteinian diffusion, i.e.

$$\tau_{rep} \sim \frac{L^2}{D}, \tag{4.188}$$

where D is the diffusion constant of the chain given by

$$D \sim \frac{k_B T}{N}. \tag{4.189}$$

You may wonder why we use $D \sim N^{-1}$ instead of $D \sim N^{-\nu}$ as in (4.165)? This is because the polymer does not move as a 'blob' of effective diameter N^ν! Instead of N^ν we decide that N is the better scaling variable. Putting the last four equations together we find

$$\tau_{rep} \sim \frac{N^3}{k_B T N_e}, \tag{4.190}$$

where we have used $\nu = 1/2$.

We understand the significance of this formula if we apply it to Fig. 4.22. Note that the horizontal separation of the red solid lines is about six orders of magnitude on the frequency axis. The ratio of the largest molecular weight to the smallest, which

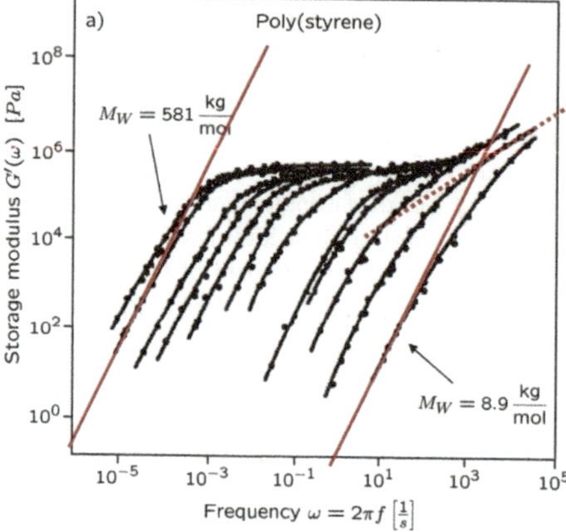

Fig. 4.22 Frequency dependence of the storage modulus of polystyrene in the molten state as function of molecular weight. This figure is identical to Fig. 4.13 except for the red lines. The solid lines possess the slope 2, whereas the dotted line has slope 2/3. These are the limiting slopes expected for $\log G'$ versus $\log \omega$ according to the Zimm model (cf. Table 4.4)

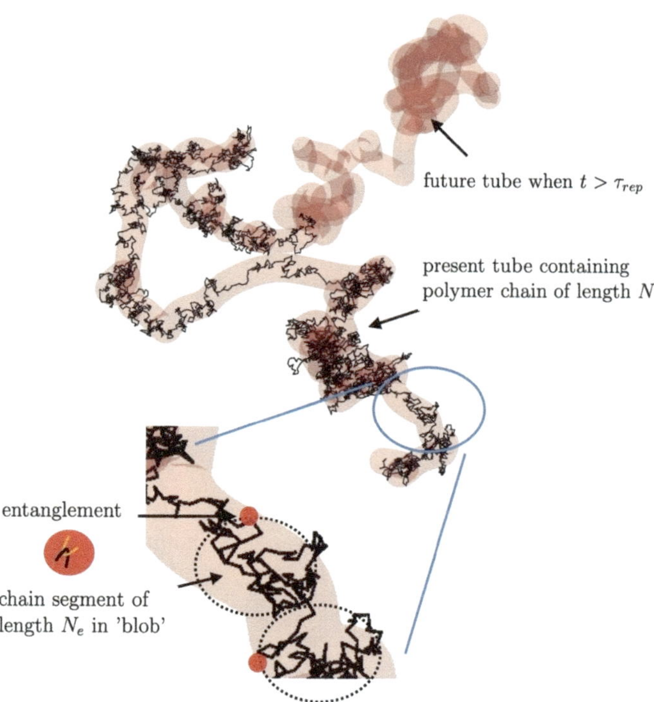

Fig. 4.23 Cartoon of a chain reptating on a path determined by entanglements with other chains or other sections of itself

is reasonably close to the molecular weight at which the G'-curves begin to show the bend which develops into the plateau, is roughly 10^2. This means that the attendant ratio of the τ_{rep}s for the two molecular weights is (again roughly) 10^6—which are just the above six decades. What (4.190) is telling us is that for times longer than τ_{rep} the chain does not 'remember' its original tube and effectively acts like the Rouse chain in the low frequency limit!

There is another quantity which we can compute to support this notion—the plateau modulus G_o, i.e. the value of G' on the plateau. We use (4.15), i.e.

$$G_o = \frac{c}{M_e} k_B T, \tag{4.191}$$

(Note that N in (4.15) is the number of chains—in the present case the number chain sections of length N_e between entanglements (or physical cross-links)!). The quantity c is the polymer mass density and $M_e \propto N_e$ is the mass of the segment of length N_e. Again we use the mass of the shortest chain in the figure, i.e. 8.9 kg mol^{-1} (The mass of the segment between two entanglements is about twice this value. But we have used the curve obtained for this mass in our above estimate and thus we stick to it here as well. After all—this is a rough calculation!). For the density we use $c = 10^3$ kg m^{-3}. The result is $G_o \approx 4 \cdot 10^5$ Pa. This is not too bad!

Remark: If we compare the storage modulus in the two panels of Fig. 4.12, we notice that the one in panel (a) behaves just as if it was one of the curves somewhere in the middle of Fig. 4.13. However, the storage modulus in panel (b) shows a plateau which apparently persists at even the smallest frequencies. In this case the chemical cross-links prevent the diffusion of the chain segments and, effectively, $\tau_{rep} = \infty$.

An even better quantity to support reptation is the translational diffusion of the chains. Here we mean spatial center of mass diffusion and not the diffusion along the tube's contour as described by (4.188). According to our above picture the tube's contour is a random path and its linear dimension R_{tube} should scale as

$$R_{tube} \sim N^{1/2}. \tag{4.192}$$

This implies for the translational center of diffusion coefficient D_{cm}

$$D_{cm} \sim \frac{R_{tube}^2}{\tau_{rep}} \sim \frac{1}{N^2}. \tag{4.193}$$

The above result of the reptation model is quite distinct from our previous results for D_{cm} based on the Rouse or the Zimm model (cf. (4.126) and (4.165)).

The current experimental status is that neither $\tau_{rep} \sim N^3$ nor $D_{cm} \sim N^{-2}$ is quite correct. Perhaps this is not too surprising. Scaling, as elegant as it may be, relies on physical intuition, which may not capture the phenomenon in its entirety. Exper-

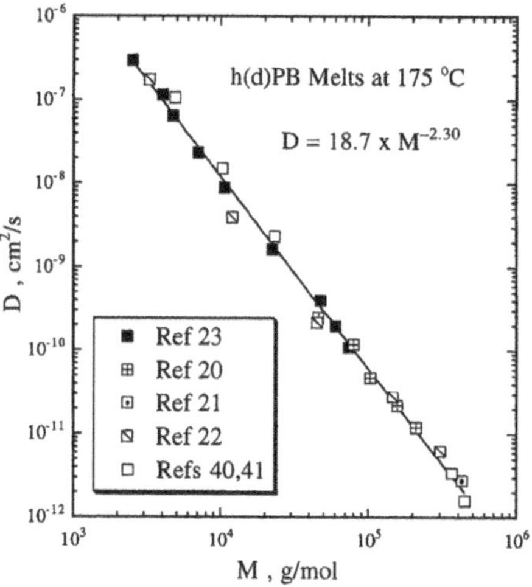

Fig. 4.24 Self-diffusion in the melt obtained from hydrogenated or deuterated polybutadiene samples adjusted to 175 °C, as a function of molecular weight. Reprinted with permission from [21]

iments appear to support exponent values which are roughly 10% larger, i.e. 3.4 instead of 3 and 2.3 instead of 2. Figure 4.24 is a figure copied from [21].

Detailed discussions/calculations related to entanglement can be found in the books written by those who invented these ideas, i.e the books by Doi and Edwards and de Gennes.

4.4 The Glass Process

In this section we discuss what is commonly called the **glass transition**. We start with a number of observation which can be made in this context. Subsequently we shall focus on one particular theoretical concept, the so called **mode coupling theory**. Even though mode coupling theory does not appear to be the ultimate answer to all observations, it yields partial answers and interesting insights despite a strongly simplified presentation.

Looking at Figs. 4.12 or 4.14 we notice that the storage modulus rises by about three orders of magnitude beyond its plateau value when the frequency becomes very high. The loss modulus on the other hand exhibits a pronounced maximum in the limit of high frequencies. Instead of changing frequency at constant temperature, we can hold the frequency constant and change the temperature. An example where this is done is shown in Fig. 4.25.

The figure shows the logarithm of the relaxation modulus for different types of polystyrene versus temperature. An atactic polymer has an irregular structure along its contour length. The structure of bulk samples of such polymers is therefore amorphous. In the case of atactic polystyrene, as we start from high temperatures, we

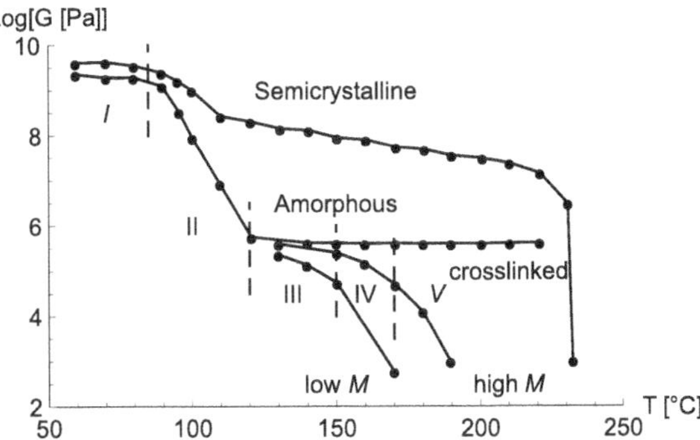

Fig. 4.25 Logarithm of the relaxation modulus (10 s) versus temperature. The polymer is isotactic (semi-crystalline) and atactic (amorphous) polystyrene. Low M and High M stand for low and high molecular weight, respectively. This figure is adopted from Fig. 1.15 in [22]

make the same observations which we already discussed in the context of Figs. 4.13 and 4.22. Depending on molecular weight, we observe a plateau of a certain width. When the polymer is chemically cross-linked, the plateau extends to the highest temperature (cf. Fig. 4.12b). Below around 120 °C the relaxation modulus rises and finally levels off about three orders of magnitude above the plateau. So, what is happening here? We have already discussed the regions labeled V, IV, and III in the case of amorphous polymers. But what is going on in the regions II and I?

A simple low-molecular weight liquid will undergo a phase transition into a crystalline phase when the temperature is continuously decreased. Can we expect something of this nature in the case of polymers? Not really. The ordering necessary to align and pack the polymer chains is entropically unfavourable. However, if the polymer's architecture is regular, e.g., it is isotactic, then ordering, at least in small spatial regions ($\sim$10 nm), can be favourable. In this case the bulk polymer matrix may be composed of crystallites surrounded by amorphous polymer.

Remark 1: If we are willing to invest work, e.g., by straining the polymer sample, this situation may be attained at temperatures when ordinarily the polymer is amorphous throughout. The process is called **strain induced crystallization** (cf. Sect. 5.1). The most prominent polymer in this context is 1,4-cis-isoprene in natural rubber.

Remark 2: The degree of crystallinity can be as high as 90% for certain low molar mass polyethylenes and as low as 5% for polyvinylchloride.

The crystallite's effect on the modulus is similar to the effect of cross-links, i.e. they increase the modulus. In Fig. 4.25 this explains the curve labeled 'Semicrystalline'. The increase in region II upon lowering the temperature is much less pronounced, since the amorphous volume fraction in the sample is smaller compared to the crystalline volume fraction, which does nothing special at this point.

All in all, regions I and II appear to be distinguished by a hard, i.e. immobile, but still amorphous polymer matrix or a certain amorphous volume fraction thereof. In region I the amorphous polymer is in a glassy state and the transition to this state in region II is called the **glass transition**. The glass transition is associated with

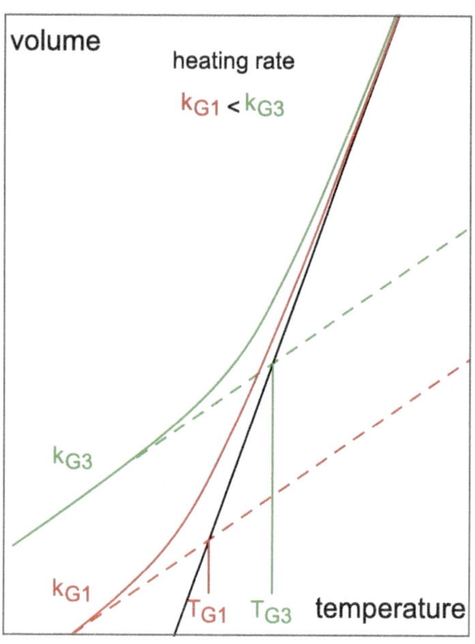

Fig. 4.26 Schematic dependence of the specific polymer volume on temperature at different heating rates. Adapted with permission from [7]

the **glass transition temperature** T_g—in general the temperature beyond which the storage modulus has levelled out. However, this is essentially a mere definition, since the glass process extends over a finite temperature range, and other features, e.g., inflection points, may be used to define T_g. In addition, T_g can be measured using different experimental methods, which lead to slightly different results.

The glass process, and therefore any value of T_g, is rate dependent. This is illustrated in the following figures. Figure 4.26 shows the specific volume versus temperature at different heating rates. Figure 4.27 shows the same for the heat capacity. Finally, Fig. 4.28 shows the dynamic moduli at different frequencies as the temperature passes through T_g. Especially the last figure, when we compare it to Fig. 4.25, tells us that we are looking at regions II and I. Figure 4.28 does not surprise us, as we do expect this behavior based on our discussion of time-temperature superposition in amorphous polymer systems. Increasing the rate or frequency shifts 'everything' towards higher temperatures. However, Fig. 4.29, which is an expanded version of Fig. 4.28, shows something new. There is another maximum of the loss modulus at a temperatures below what we just defined as T_g. This is another relaxation process.

Why can we tell that this is a relaxation process? Remember our calculation of (4.135) and (4.136) based on the relaxation modulus (4.134). When we consider a single τ'_p only, we obtain (4.135) and (4.136) without the summation—corresponding to a single step and a single maximum. In other words, different slopes of the relaxation modulus along the time axis, signalling different processes, translate into corresponding features in G' and G''.

In particular, the two maxima of the loss modulus exhibit different shifts when different frequencies are compared. Figure 4.30 shows this explicitly. The first max-

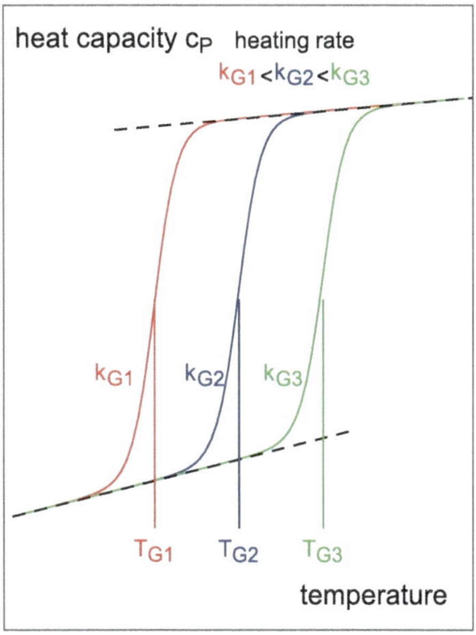

Fig. 4.27 Schematic dependence of the isobaric polymer heat capacity on temperature at different heating rates. Adapted with permission from [7]

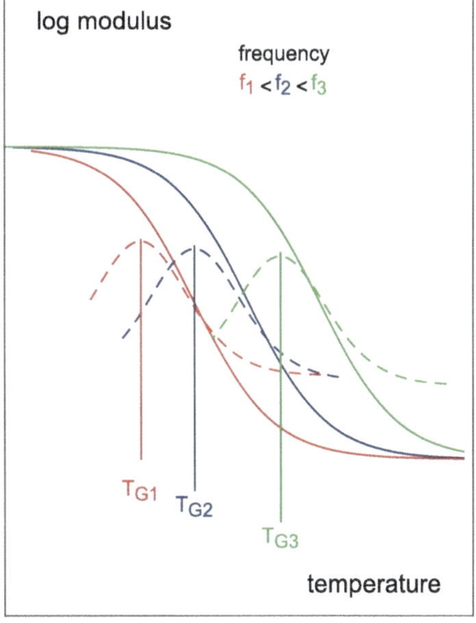

Fig. 4.28 Schematic dependence of the dynamic moduli (solid lines: G'; dashed lines: G'') on temperature at different frequencies. Adapted with permission from [7]

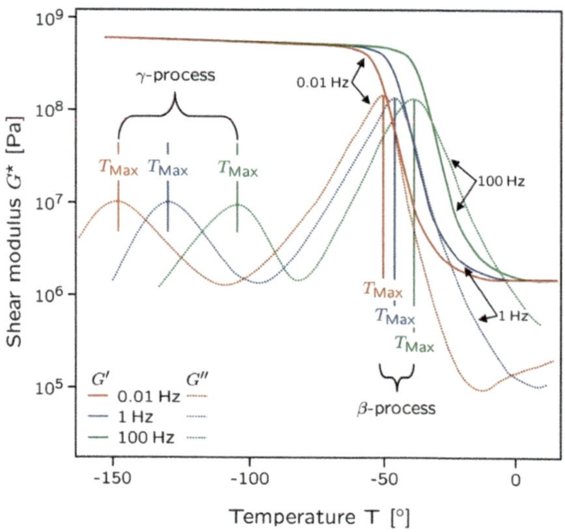

Fig. 4.29 Extended version of Fig. 4.28 revealing two different relaxation processes. The polymer is a solution styrene-butadiene rubber (S-SBR). Reprinted with permission from Fig. 2.47 in [7]

Fig. 4.30 Frequency versus inverse temperature for the processes denoted β- and γ-processes in Fig. 4.29. Reprinted with permission from Fig. 2.47 in [7]

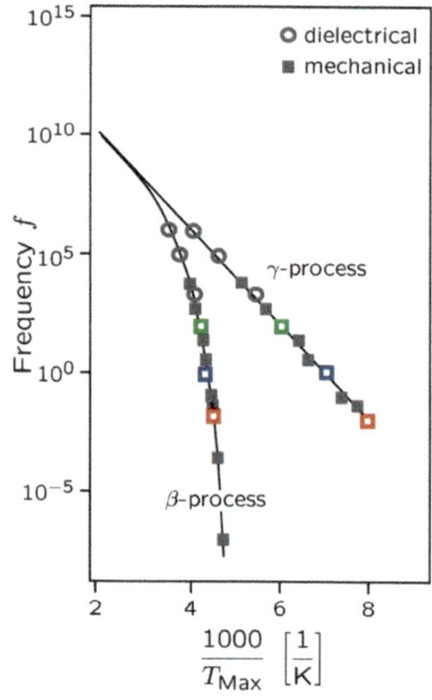

imum, i.e. the maximum at or close to T_g, follows the curve labeled 'β-process' and the second maximum, the one occurring at a lower temperature, follows the curve labeled 'γ-process'. Based on our previous discussion of time-temperature superposition our expectation is that there should only be one curve.

A Remark on notation: It is customary to assign greek letters, α, β, γ, δ,... to relaxation processes in their order of appearance as temperature decreases. The α process is customary the glass transition. Here the sequence starts with β, however.

The curve labeled 'β-process' in Fig. 4.30 can be described quite well using (4.73), where a_T is given by (4.72) and the reference temperature is taken to be T_g for a particular ω_g, i.e.

$$\ln \omega = \ln \omega_g + c_1 \frac{T - T_g}{c_2 + T - T_g}, \tag{4.194}$$

where

$$c_1 = A/v_f(T_g) \quad \text{and} \quad c_2 = v_f(T_g)/\alpha. \tag{4.195}$$

Note that in the literature (4.194) can be found with log instead of ln. In this case c_1 differs from c_1 in (4.194) by the factor $\log e \approx 0.43$.

Before we discuss the application of this fit function to the data in Fig. 4.30, we want to mention an equivalent (empirical) form of the Dolittle relation (4.69) (after it is combined with (4.71)), which is meant to specifically describe the viscosity at or near T_g, i.e.

$$\eta(T) = B \exp\left[T_a/(T - T_{VF})\right]. \tag{4.196}$$

Equation (4.196) is called the **Vogel–Fulcher law** and T_{VF} is the Vogel or **Vogel–Fulcher temperature**. The temperature T_a is an activation temperature, whereas B is a constant independent of temperature. T_{VF} and T_a are related to c_1 and c_2 in (4.194) via

$$T_{VF} = T_g - v_f(T_g)/\alpha \equiv T_g - c_2 \tag{4.197}$$

and

$$T_a = A/\alpha \equiv c_1 c_2 \tag{4.198}$$

(check this!). Multiplied with the gas constant R this activation temperature yields an activation energy $E_a = RT_a$. Why there should be an activated process involved here becomes plausible when we look at the curve labeled 'γ-process' in Fig. 4.30. This curve, apparently a straight line, appears to coincides with the curved line based on (4.194) when the temperature is sufficiently high.

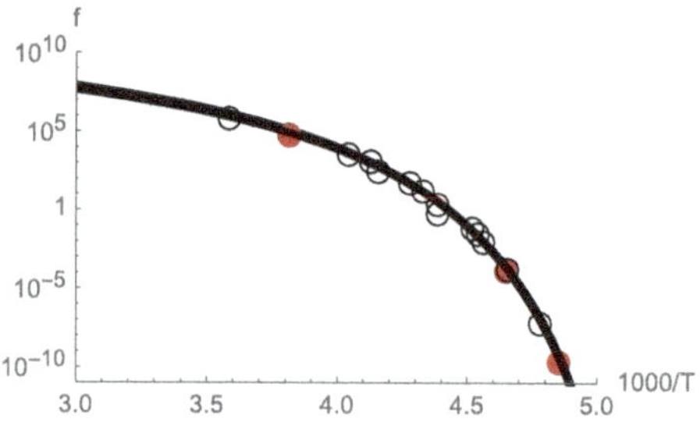

Fig. 4.31 The data points labeled β-process in Fig. 4.30 fitted with (4.194) using three distinct pairs (ω_g, T_g) (red circles) along the curve mapped out by the data points

The fact that the maxima of the 'γ-process' appear to fall onto a straight line suggests that this process can be described by an **Arrhenius equation**:

$$\ln k = \ln A_a - \frac{E_a/R}{T}. \tag{4.199}$$

Here A_a is a factor and E_a is the **activation energy** in mole of the process. The rate constant k gives the frequency of the process. If we expand (4.194) in the limit $T \gg T_g$, we find

$$\ln \omega = \ln \omega_g + \frac{A}{v_f(T_g)} - \frac{A/\alpha}{T} \tag{4.200}$$

to leading order in T^{-1}. Since, as already mentioned, the data points for both processes seem to come together at high temperatures, we may concluded that T_a from (4.198) is indeed the same as E_a/R in (4.199). This means that both processes should be closely related, if not identical, in this temperature limit.

Figure 4.31 shows the data points labeled β-process in Fig. 4.30 fitted using (4.194). First we select a point along the curve mapped out by the data in Fig. 4.30, which defines the reference $\omega_g = 2\pi f_g$ and the reference T_g in (4.194). For the pair (ω_g, T_g) we attempt the best fit through the data points in Fig. 4.30 by adjusting c_1 and c_2. Here this is done for three pairs (ω_g, T_g) indicated by the red circles. The resulting values of c_1 and c_2 are compiled in Table 4.5. What the table also shows is that no matter which pair (ω_g, T_g) we use in (4.194), we always get the same T_{VF} and E_a. The Vogel–Fulcher temperature T_{VF} is T in the limit $\omega \to 0$ in (4.194), i.e. T_{VF} is the limiting T_g for an infinitely slow process.[8] The value for E_a is what one expects for interactions between neighboring polymer chain segments containing only a few

---

[8] For those of you who compare and check values: The data in this example are taken from Fig. 2.47 in [7]. The polymer is an S-SBR. In Table 2.2 in the same reference T_{VG} for S-SBR is $-60\,^\circ$C, i.e. 30$\,^\circ$C higher in comparison to the value in Table 4.5. However, T_{VF} depends critically on the styrene to vinyl ratio, which in the case of $T_{VG} = -60\,^\circ$C is 25% styrene to 50% vinyl. In other words, it appears that the styrene content in the S-SBR in Fig. 2.47 is lower.

Table 4.5 Application of (4.194)

f_g [Hz]	T_g [K]	c_1	c_2 [K]	$T_{VF} =$ $T_g - c_2$ [K]	$E_a = Rc_1c_2$ [kJ/mol]
$10^{-9.5}$	206	47	23	183	9.0
$10^{-3.7}$	215	34	32	183	9.0
$8.4 \cdot 10^4$	262	13.3	79	183	8.7

carbon atoms. Hence, while T_g depends on frequency or the rate of the process, it is possible to obtain a meaningful limiting T_g via T_{VF}. In addition, we obtain E_a, providing additional information about the underlying molecular processes. When we discuss polymers containing fillers, we shall return to E_a and learn that the observed E_a can be due to processes which have nothing to do with the polymer itself (cf. [23]).

Our phenomenological description of the glass process thus far has heavily relied on free volume as a central ingredient. It turns out that free volume is also useful when we want to understand how T_g is affected by molecular weight or molecular architecture in general. Here we avoid a general discussion though. Instead we focus on the molecular weight dependence of T_g for linear polymers.

We consider a constant volume V containing N polymers containing M monomers each. Equation (4.71) ties the free volume $v_f(T)$ in this system, at temperature T, to the glass transition temperature T_g if it is our T_r. Hence can rewrite (4.71) as

$$v_f(T) = v_f(T_{g,\infty}) + \delta v_f(M) + \alpha(T - T_g). \tag{4.201}$$

This means that $v_f(T_g)$ is divided into the free volume at $T_{g,\infty}$ when M is infinite, i.e. there are no free ends inside V, plus a piece $\delta v_f(M)$, which accounts for the extra free volume due to the presence of the free ends when M is finite. Setting $T = T_{g,\infty}$ in (4.201) yields

$$T_g = T_{g,\infty} - \frac{\delta v_f(M)}{\alpha}. \tag{4.202}$$

We also assume $\delta v_f(M) \propto 2N$, where $2N$ is the number of polymer ends. Since $NM/V = \rho$, where ρ is the constant number density of monomers, we obtain

$$T_g = T_{g,\infty} - \frac{c}{M}, \tag{4.203}$$

where c is a quantity independent of M. In other words, decreasing M should decrease the glass transition temperature. This relation between the glass transition temperature and the polymer molecular weight was first discussed by Fox and Flory [24] (Fig. 3 in this paper shows a nice example confirming (4.203)).

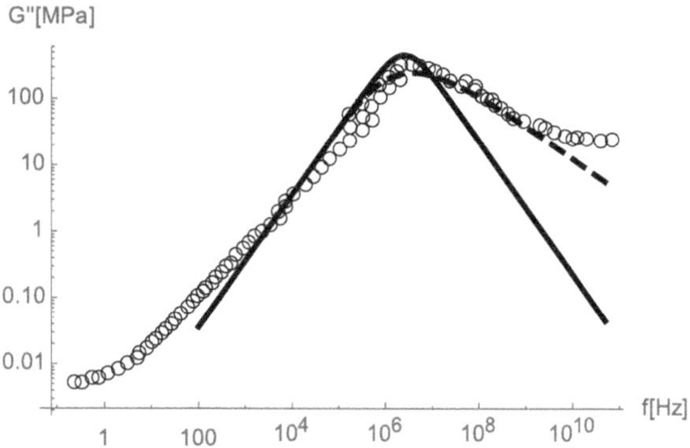

Fig. 4.32 Experimental loss modulus of highly cross-linked polyisoprene compared to $G''(\omega)$ ($\omega = 2\pi f$) calculated from the exponential (solid line) and a stretched exponential form (dashed line) of the relaxation modulus (data reproduced with the permission of Continental Reifen Deutschland)

There is another observation to be made. This one concerns the relaxation modulus $G(t)$ near T_g. The exponential form

$$G(t) \sim \exp[-t/\tau] \tag{4.204}$$

yields

$$G''(\omega) \sim \frac{\omega\tau}{(\omega\tau)^2 + 1} \tag{4.205}$$

as we had found discussing the models of Rouse and Zimm. The solid line in Fig. 4.32 is a fit based on this form to data (open circles) obtained for the loss modulus of a sample of highly cross-linked polyisoprene. We notice that the experimental peak, corresponding to what we called the β-process, in this log-log plot is quite asymmetric, whereas the exponential $G(t)$ in (4.204) produces a symmetric peak. Also included in the figure is a dashed line, which is a much better fit capturing the aforementioned asymmetry. The dashed line is calculated with a **Kohlrausch–Williams–Watts** or **stretched exponential** form of the relaxation modulus, i.e.

$$G(t) \sim \exp[-(t/\tau)^{\beta}], \tag{4.206}$$

where β is an exponent—usually between 0.3 and 0.5 (here $\beta = 0.5$). Since the attendant $G''(\omega)$ can only be presented as an unwieldy expression involving special functions, we do not give it here.

Remark: If $G(t)$ has a power-law form, i.e.

$$G(t) \sim t^{-\beta}, \tag{4.207}$$

then

$$G'(\omega), G''(\omega) \sim \omega^{\beta}, \tag{4.208}$$

respectively.

Nevertheless, despite all of this, the inescapable conclusion thus far is the following: Close to T_g and below we find relaxation processes of different nature for which, at this point, we have no microscopic theory. Since free volume has been a key ingredient in our discussion of the observations in conjunction with the glass process (remember in particular our prediction of how T_g depends on molecular weight), one might try to base a theory of the glass transition on free volume (in fact, the free volume based description of dense gases and liquids dates back to the 1930s (see [25])). There are many theoretical models, including free volume theories, attempting to describe the glass transition and here we cannot do justice to even the more important ones. Instead we want to discuss the so called **mode coupling theory** (MCT)—albeit in a much abbreviated and idealized form.

MCT is not a polymer theory per se. Rather, it is a theory for the glass process in general, i.e. the reversible process during which liquid-like amorphous order is replaced by solid-like amorphous order and vice-versa. A process which occurs in low molecular weight systems as well. Neither does MCT describes all our observations. But MCT features a number of interesting and thought-provoking ingredients and results, which a student of polymer physics might want to now about. Most of the following is borrowed from an article by Janssen [26]. A discussion of the glass transition in polymer systems, including a number of theoretical approaches, can be found, for example, in [27].

Standard MCT seeks to predict the dynamics of the function $F(q, t)$, a time-dependent structure factor, given by

$$F(q, t) = \frac{1}{N} \langle \hat{\rho}(-\vec{q}, 0)\hat{\rho}(\vec{q}, t) \rangle \tag{4.209}$$

where

$$\hat{\rho}(\vec{q}, t) = \int d^3 r \, e^{i\vec{q}\cdot\vec{r}} \rho(\vec{r}, t) \tag{4.210}$$

and

$$\rho(\vec{r}, t) = \sum_{i=1}^{N} \delta(\vec{r} - \vec{r}_i(t)). \tag{4.211}$$

Note that $\rho(\vec{r}, t)$ is the same as in (2.97), except that this is not specifically a monomer number density but a general particle density. Hence $F(q, t)$ is very much like $P_i(\theta)$, except that now the time dependence is included.

We shall not derive the equation governing the dynamic evolution of $F(q, t)$. This calculation can be found in the references in the aforementioned article by Janssen. Instead we want to study this equation in an idealized approximation, i.e.

$$\ddot{x}(t) + \omega^2 x(t) + a \int_0^t x^2(t - t')\dot{x}(t')dt' = 0. \tag{4.212}$$

Here $x(t)$ stands for $F(q, t)/F(q, 0)$ and in particular $x(t = 0) = 1$. In (4.212) the q-dependence is ignored and, in addition, $\omega \ (> 0)$ and $a \ (> 0)$ are functions, which here are treated as constants. Despite this, (4.212) captures most of the essence of the full MCT equation for $F(q, t)$.

Note that (4.212) becomes the equation of motion of a one-dimensional damped harmonic oscillator if we replace $x^2(t - t')$ with $\delta(t - t')$, i.e. $\ddot{x}(t) + \omega^2 x(t) + a\dot{x}(t) = 0$. We can solve (4.212) numerically using a **Verlet algorithm**. First we expand $x(t)$, i.e.

$$x(t \pm \delta t) = x(t) \pm \dot{x}(t)\delta t + \frac{1}{2}\ddot{x}(t)\delta t^2 + \cdots . \tag{4.213}$$

Subsequently we add the two equations $(\pm!)$ and insert $\ddot{x}(t)$ from (4.212), i.e.

$$x_{n+1} \approx 2x_n - x_{n-1} - \left[\omega^2 x_n + a \sum_{i=0}^{n-1} x_{n-i}^2 (x_{i+1} - x_i)\right]\delta t^2. \tag{4.214}$$

Note that $x(t = n\delta t) \equiv x_n$. In addition, we approximate the integral by a summation. Our starting x-values are $x_0 = 1$ and $x_1 \approx 1 - (\omega^2/2)\delta t + (a\omega^2/6)\delta t^3$. The latter we obtain by inserting a power series expansion of $x(t)$ into the above damped harmonic oscillator-approximation of (4.212). Straightforward iteration of (4.214) yields the graphs depicted in Fig. 4.33. Compared to the simple damped harmonic oscillator the solutions of the idealized MCT, i.e. (4.212), exhibit an increasing slowdown of the decay of $x(t)$ as a is increased. The slowdown of the decay of $x(t)$ is the result of the non-linearity of (4.212). Careful analysis reveals that the slowdown becomes infinite for $a \geq 4$. With regard to $F(q, t)$ this means that the system becomes non-ergodic for $a \geq 4$.

In order to better understand what is going on, we digress from the glass process and briefly turn to chemical reactions—or more precisely to rate equations. Equation (4.215) describes how the mole fraction of the chemical component X changes with time during a reaction. Specifically this is a non-linear autocatalytic reaction due to the term proportional to n_X^2 on the right hand side of the equation. $n_Y(0)$ is the mole fraction of chemical Y, which is constant and the various ks are rate constants. Equation (4.215) is one equation within a larger reaction schema discussed in Sect. 7.3.1 of [28]:

$$\frac{d}{dt}n_X = (k_1 n_Y(0) - k_3)n_X - k_{-1}n_X^2. \tag{4.215}$$

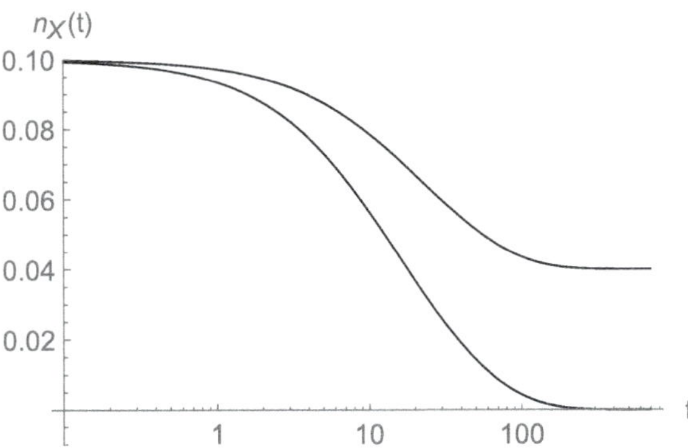

Fig. 4.33 Numerical solutions of (4.212) for $\omega^2 = 1$ and for $a = 3.9$ (blue) in comparison to $a = 4.0$ (golden). The dotted curves are the corresponding solutions of the damped harmonic oscillator, i.e. $\ddot{x}(t) + \omega^2 x(t) + a\dot{x}(t) = 0$

Fig. 4.34 Stable solutions of (4.215) for $n_Y(0)$ slightly below $n_{Y,crit}$ in comparison to $n_Y(0)$ slightly above $n_{Y,crit}$. Here $n_{Y,crit} = 0.5$. $n_Y(0) = 0.49$ (lower curve) and $n_Y(0) = 0.51$ (upper curve); $k_{-1} = 0.5$, $n_X(0) = 0.1$, $k_3 = 1$, and $k_1 = 2$

Integration of this equation yields the solution

$$n_X(t) = \frac{(k_1 n_Y(0) - k_3)n_X(0)}{k_{-1}n_X(0) - [k_{-1}n_X(0) - k_1 n_Y(0) + k_3]\exp[(-k_1 n_Y(0) + k_3)t]}.$$

Depending on whether $n_Y(0) < n_{Y,crit}$ or $n_Y(0) > n_{Y,crit}$, where $n_{Y,crit} = k_3/k_1$, there are two steady state solutions $n_X(\infty) = 0$ or $n_X(\infty) = (k_1 n_Y(0) - k_3)/k_{-1}$ if $n_X(0) > 0$. The system's choice which of the two solutions it prefers, i.e. the stable solution, depends on the parameter $n_Y(0)/n_{Y,crit}$. Figure 4.34 illustrates this showing the stable solutions for $n_Y(0)$ slightly below in comparison to $n_Y(0)$ slightly above $n_{Y,crit}$. Autocatalytic reactions are interesting because of the possible bifurcation of the long time concentration of chemical components, depending on the value of a control parameter. This mechanism is believed to be important during **chemical evolution**, i.e. without violation of the second law nature is able to produce increasingly complex systems.

Comparing (4.212) with (4.215) we note that a/a_c ($a_c = 4$) in the former equation is akin to $n_Y(0)/n_{Y,crit}$ in the latter equation. The difference, however, is that $x(t)$ exhibits a delayed decay. This has to do with the memory kernel in the integral, which links t to all previous times.

Figure 4.35 is a schematic summary of the full MCT result (note that $S(q) = F(q, 0)$) obtained for particle systems. The central element is a cage around each

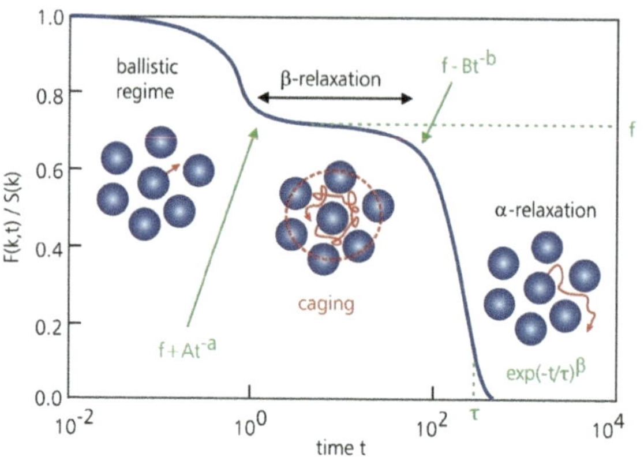

Fig. 4.35 Schematic summary of the full MCT result. Reprinted from [26] (copyright ©2018 Janssen)

particle. At very short times the particle moves more or less without noticing the other particles forming the cage. This is the **ballistic regime**. In our idealized MCT this time regime is dominated by the harmonic potential. In an intermediate time regime (here the β-relaxation), which grows as T approaches T_c (in the idealized MCT this means a approaches $a_c = 4$; in the full MCT the factor corresponding to a is a function linearly dependent on temperature.), the cage restrains the particle's motion. Both limits of this regime are characterized by power law behavior. Finally, when $T < T_c$, the particle escapes the cage (here α-relaxation). This ultimate decay of $F(q, t)/F(q, 0)$, characterized by the relaxation time τ, follows a stretched exponential.

But how does this relate to our observations during the glass process? Within MCT T_c is the glass transition temperature and one finds that $\tau \sim (T - T_c)^{-\gamma}$ ($\gamma > 0$). With $\tau \propto \eta$, this power law of course differs from the exponential forms of the Vogel–Fulcher law (4.196) or the Dolittle relation (4.69) in conjunction with the linear temperature dependence of the free volume (4.71). On the other hand, MCT predicts that $F(q, t) = \tilde{F}(q, t/\tau(T))$, where $\tilde{F}(q, t/\tau(T))$ is a master function. This corresponds to time-temperature superposition! And of course there is the stretched exponential decay, during the α-relaxation in MCT. In addition, other predictions of MCT have been confirmed. But since these do not pertain to our above observations we shall not discuss them here.

4.5　Problems

Problem 4.1 In this problem we estimate the **compression modulus** K of elastomers. Our starting point is the thermodynamic equation $dE = -PdV + TdS$. We assume that $T = 0$. Hence we now have $dU = -PdV$ instead, where the potential energy is given by

$$U \approx \sum_{i<j}^{N} u_{ij}.$$

(4.216)

Model the pair interaction u_{ij} via the **Lennard–Jones potential**

$$u_{ij} = 4\epsilon \left[\left(\frac{\sigma}{r_{ij}} \right)^{12} - \left(\frac{\sigma}{r_{ij}} \right)^{6} \right].$$

(4.217)

The indexes i and j stand for suitable groups of atoms (called particles in the following), e.g. a monomer unit, separated by the distance r_{ij}. N is the number of particles. The quantities ϵ and σ are parameters.

(a) Rewrite the summation in (4.216) as an approximate integration from r to ∞. Here r is the closest separation between two particles. Solve the integral and express U/N in terms of ϵ, σ, and $\rho = 1/r^3$, where ρ is the particle number density.

(b) Find the equilibrium density ρ_o in terms of σ via $d(U/N)/d\rho|_{\rho_o} = 0$.

(c) Find K in terms of ϵ and σ.

(d) Calculate K using the parameter values $\epsilon/k_B = 148.7\,\mathrm{K}$ and $\sigma = 3.79\,\mathrm{\AA}$. Remark: These values were obtained by fitting the second virial coefficient of methane via the Lennard–Jones potential. Even though methane is not a monomer unit, it proves to be a reasonably suitable molecule in the present context. Compare your result for K to the values published in [30]. Another remark: A keen observer may notice the discrepancy between this ϵ-value and the one we used before in (2.42). This is because (2.42) is merely a correction to the torsion potential, whereas here the Lennard–Jones potential describes the entire interaction between two particles.

Problem 4.2 (a) Based on the data in Fig. 4.3 provide a rough estimate of the segment length, i.e. the average number of monomers between two cross-links. Use $920\,\mathrm{kg\,m^{-3}}$ for the mass density of natural rubber.

(b) The cross-linking agent in this case is dicumyl peroxide. Its molecular weight is $M = 270.4\,\mathrm{g/mol}$. Assuming that every dicumyl peroxide molecule produces one cross-link, calculate the segment length if 1.5% by weight are added to and react with the polymer.

Problem 4.3 (a) Consider (4.76) in the case of steady shear, i.e. $\dot{\gamma}(t) = \dot{\gamma} = const$, and show that the **viscosity coefficient** is given by

$$\eta = \int_0^{\infty} G(t)dt,$$

(4.218)

where $G(t)$ is the **shear relaxation modulus**.

(b) Apply this equation to the **Maxwell model** and compare your result to (A.6).

Problem 4.4 In this problem we want to calculate the N-dependence of the **intrinsic viscosity** η_i of a polymer solution defined via

$$\eta_i = \lim_{c \to 0} \frac{\eta_{solution} - \eta_s}{\eta_s \, c}. \tag{4.219}$$

Here $\eta_{solution}$ is the viscosity of the solution, η_s is the viscosity of the pure solvent, and c is the polymer mass concentration.

(a) Show that (4.218) implies

$$\eta \sim G(\tau)\tau. \tag{4.220}$$

Assume that $G(t) \sim A \exp[-t/\tau]$, where A is a constant.

(b) Express $G(\tau)$ in terms of $k_B T$ and the number density of chains ρ and obtain a relation expressing η_i in terms of N, i.e. the number of monomers (or alternatively Kuhn segments), whose volume is b^3, and τ. Note: $\eta = \eta_{solution} - \eta_s$.

(c) Define the time τ_R via the Einstein diffusion relation for the polymer center of mass

$$D_{cm} = R^2/\tau_R, \tag{4.221}$$

where R is the linear dimension of the polymer chain in solution. Eliminate D_{cm} from this equation using our previous result for the center of mass diffusion according to the Rouse model. What is the resulting dependence of τ_R, now called the **Rouse time**, on N?

(d) By equating τ in (4.220) with τ_R obtain the N-dependence of the intrinsic viscosity predicted by the Rouse model.

(e) What is the analogous prediction of the Zimm model?

(f) Plot the data, molar mass versus intrinsic velocity, from Table I in [29] for polystyrene in benzene (T $= 25\,°$C) and cyclohexane (T $= 34.5\,°$C). What are the respective slopes of straight-line fits through the data in a log-log plot? How do these values compare to the theoretical predictions?

Problem 4.5 (a) Derive (A.7) for the Zener model.

(b)   Derive the dynamic moduli and the loss tangent, i.e. derive the (A.8) to (A.10).

References

1. A. Blume, J. Kiesewetter, Determination of the crosslink density of tire tread compounds by different analytical methods. Kautschuk Gummi Kunststoffe **9**, 33 (2019)
2. L.R.G. Treloar, Stress-strain data for Vulcanised natural rubber under various types of deformation. Trans. Faraday Soc. **40**, 59 (1944)
3. M.C. Shen, D.A. McQuarrie, J.L. Jackson, Thermoelastic behavior of natural rubber. J. Appl. Phys. **38**, 791 (1967)
4. R. Hentschke, Including temperature effects in the theory and simulation of problems in rubber reinforcement, in *Advances in Understanding Thermal Effects in Rubber* ed. by G. Heinrich et al., Advances in Polymer Science 294. (Springer, Heidelberg, 2024)
5. C. Ortiz, G. Hadziioannou, Entropic elasticity of single polymer chains of poly(methacrylic acid) measured by atomic force microscopy. Macromolecules **32**, 780 (1999)
6. A.G. Marangoni, L.H. Wesdorp, *Structure and Properties of Fat Crystal Networks*, 2nd edn. (CRC Press, Boca Raton, 2013)
7. C. Wrana, *Introduction to Polymer Physics* (LanXESS, Leverkusen, 2009)
8. S. Onogi, T. Masuda, K. Kitagawa, Rheological properties of anionic polystyrenes. I. Dynamic viscoelasticity of narrow-distribution polystyrenes. Macromolecules **3**, 109 (1970)
9. A. Lang, Experimentelle und theoretische Untersuchungen zum Reibverhalten elastomerer Werkstoffe auf rauen Oberflächen. Ph.D. Thesis, University of Hannover (2018)
10. J.D. Ferry, *Viscoelastic Properties of Polymers* (Wiley, New York, 1961)
11. M.L. Williams, J.D. Ferry, Second approximation calculations of mechanical and electrical relaxation and retardation distributions. J. Polymer Sci. **11**, 169 (1953)
12. M.L. Williams, R.F. Landel, J.D. Ferry, The temperature dependence of relaxation mechanisms in amorphous polymers and other glass-forming liquids. J. Am. Chem. Soc. **77**, 3701 (1955)
13. K.A. Grosch, The rolling resistance, wear and traction properties of tread compounds. Rubber Chem. Technol. **69**, 495 (1996)
14. B. Lorenz, Y.R. Oh, S.K. Nam, S.H. Jeon, B.N.J. Persson, Rubber friction on road surfaces: experiment and theory for low sliding speeds. J. Chem. Phys. **142**, 194701 (2015)
15. P.E. Rouse, A theory of the linear viscoelastic properties of dilute solutions of coiling polymers. J. Chem. Phys. **21**, 1272 (1953)
16. B.H. Zimm, Dynamics of polymer molecules in dilute solution: viscoelasticity, flow birefringence and dielectric loss. J. Chem. Phys. **24**, 269 (1956)
17. M. Abramowitz, I.A. Stegun, ed. by *Handbook of Mathematical Functions* (Dover, New York, 1972)
18. R. Hentschke, *Statistische Mechanik* (Wiley-VCH, Weinheim, 2004)
19. D.F. Hodgson, E.J. Amis, Dynamic viscoelasticity of dilute polyelectrolyte solutions. J. Chem. Phys. **94**, 4581 (1991)
20. R. Johnson, J. Schrag, J. Ferry, Infinite-dilution viscoelastic properties of polystyrene in θ-solvents and good solvents. Polymer Jpn. **1**, 742 (1970)
21. T.P. Lodge, Reconciliation of the molecular weight dependence of diffusion and viscosity in entangled polymers. Phys. Rev. Lett. **83**, 3218 (1999)
22. U.W. Gedde, *Polymer Physics* (Chapman & Hall, London, 1995)
23. J. Fritzsche, M. Klüppel, Structural dynamics and interfacial properties of filler-reinforced elastomers. J. Phys.: Condensed Matter **23**, 035104 (2011)
24. T.G. Fox, P.J. Flory, Second-order transition temperatures and related properties of polystyrene. I. Influence of molecular weight. J. Appl. Phys. **21**, 581 (1950)
25. J.O. Hirschfelder, C.F. Curtiss, R.B. Bird, *Molecular Theory of Gases and Liquids* (Wiley, New York, 1954)
26. L.M.C. Janssen, Mode-coupling theory of the glass transition: a primer. Front. Phys. **6**, Article 97 (2018)
27. I.M. Kalogeras, Glass-transition phenomena in polymer blends, in *Encyclopedia of Polymer Blends: Volume 3: Structure*, ed. by A.I. Isayev, 1st edn. (Wiley-VCH, Weinheim, 2016)

28. R. Hentschke, *Thermodynamics*, 2nd edn. (Springer, Heidelberg, 2022)
29. Y. Einaga, Y. Miyaki, H. Fujita, Intrinsic viscosity of polystyrene. J. Polym. Sci. Polym. Phys. **17**, 2103 (1979)
30. J. Burns, P.S. Dubbelday, R.Y. Ting, Dynamic bulk modulus of various elastomers. J. Poly. Sci. **28**, 1187 (1990)

Chapter 5
Selected Topics

Abstract The subsequent collection of 'selected topics' offers a mere glimpse of what the respective headings announce. This is because each topic by itself may fill an entire book or even several books. In turn this means that what you find under each heading is biased by this author's personal experience with the subject and therefore may neglect material considered essential by others. For this I apologize. However, I have included references which most likely cover the omitted material. Specifically, we shall discuss the mechanics of polymers, filler effects, albeit mainly in the context of elastomers, and some theoretical concepts in the areas of polymer liquid crystals and polyelectrolytes.

5.1 Aspects of the Mechanics of Polymers

● *Basic Stress-Strain Curves*

The measurement of (tensile) stress-strain curves ranks among the most common mechanical tests carried out on polymer samples. However, there are different types of stress. **True stress**, for instance, is the applied load divided by the actual sample cross-sectional area. **Engineering stress** or **nominal stress**, on the other hand, is the applied load divided by the original cross-sectional area of the sample.

Figure 5.1 shows schematic stress-strain curves for different types of polymers. Polymer single crystal fibers show the steepest increase of stress with increasing strain. This is because the covalent bonds of the polymer backbone usually are highly aligned with the drawing direction. How this is done we shall briefly discuss in the section on liquid crystalline polymers. Since covalent bonds are very strong, the stress at break of these polymers is quite high (in the GPa range). On the other hand, the strain at break is only a few %.

Figure 5.2 shows an example of the fracture strength of polymer single crystal fibers as a function of their diameter. The solid line fitting the data quite well is simply based on $\sigma_T \propto D^{-1}$, where σ_T is the **tensile strength** and D is the fiber diameter. This proportionality is motivated by the idea that the initial 'damage' or cause leading to the failure of the fiber originates at its surface. The observation that

© The Author(s), under exclusive license to Springer Nature Switzerland AG 2025 151
R. Hentschke, *A Concise Introduction to Polymer Physics*, Undergraduate Lecture Notes in Physics, https://doi.org/10.1007/978-3-031-87324-9_5

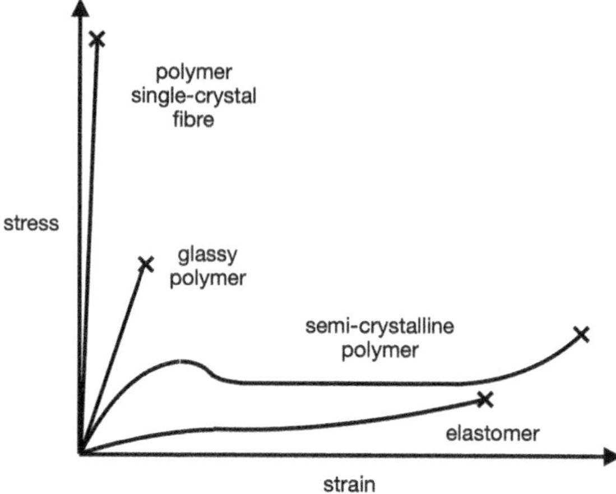

Fig. 5.1 Schematic stress-strain curves of the indicated polymer types. Crosses indicate the end of the respective curve due to sample failure

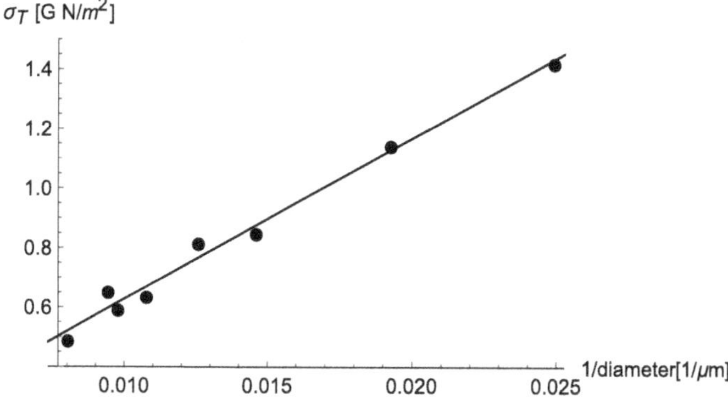

Fig. 5.2 Tensile strength σ_T of polydiacetylene single crystal fibres as function of their inverse diameter. The data points are from Fig. 5.64 in [2]. The solid line is a fit using $\sigma_T \propto (\text{diameter})^{-1}$

'thinner is stronger' actually is quite general in the context of technical as well as natural polymer fibers. Another quantity important for the tensile strength of fibers is the molecular weight of the polymers. Higher molecular weight generally means increased tensile strength [1].

The interesting shape of the stress-strain curve of semi-crystalline polymers is due to the complex structure consisting of amorphous and crystalline regions. On the molecular level the crystalline regions restructure by various mechanisms (slip, twinning,...) in the different strain regions. The stress-strain curve of elastomers is discussed in various places and in some detail in these notes.

Sample failure, indicated by the crosses, is a difficult topic. This is because failure and where it happens along the stress-strain curve, is influenced by diverse structural

and dynamic effects. Here we do not want to discuss this topic and refer the reader
to the literature dedicated to it (e.g., [3, 4]).

● *Mooney–Rivlin Theory*

In Sect. 3.5 we derived the elastic free energy contribution due to the macroscopic
deformation of a polymer bulk sample expressed in terms of the quantities λ_x, λ_y, and
λ_z. Each of these quantities is a factor by which the attendant edge of a rectangular
volume is compressed or stretched during a deformation of the original rectangular
volume.

Mooney [5] and Rivlin [6] had the idea to derive a free energy of deformation
expressed in terms and valid for arbitrary λ_α based on symmetry arguments. The
upshot is that the following expressions, quadratic in the λs and valid for any type
of deformation, do not change if we change coordinate systems:

$$I_1 \equiv \lambda_x{}^2 + \lambda_y{}^2 + \lambda_z{}^2 \tag{5.1}$$
$$I_2 \equiv \lambda_x{}^2\lambda_y{}^2 + \lambda_x{}^2\lambda_z{}^2 + \lambda_y{}^2\lambda_z{}^2 \tag{5.2}$$
$$I_3 \equiv \lambda_x{}^2\lambda_y{}^2\lambda_z{}^2 \tag{5.3}$$

These invariants appear most clearly in Rivlins article (cf. his equation 3.4). Essentially, he uses the fact that the strain tensor can be diagonalized at every point in the
elastic body. The λ_α are the stretch factors associated with the orthogonal axes (principal axes of strain) in the coordinate system in which the strain tensor is diagonal
(Remark 1: The λ_α are the same as the factors $(1 + u^{(1)})$ in (4.7); Remark 2: In the
case of uniform stretch of a square column the principal axes and the identification
of the λs is rather obvious. In the case of shear this is much less so!). Note that I_1
as well as (in principle) I_3 appear in the elastic entropy (3.51). I_2 is new and if we
include it we may construct an elastic free energy of the following form:

$$\Delta F^{el} = C_1(I_1 - 3) + C_2(I_2 - 3) + C_3(I_3 - 3). \tag{5.4}$$

Here C_1, C_2, and C_3 are as yet unknown constants. Note also that $\Delta F^{el} = 0$ if there
is no deformation.

In the case of uniaxial stretch at constant volume, i.e. $I_3 = 0$, (cf. (3.56) and
(4.20)) ΔF^{el} becomes

$$\Delta F^{el} = C_1\left(\lambda^2 + \frac{2}{\lambda} - 3\right) + C_2\left(2\lambda + \frac{1}{\lambda^2} - 3\right). \tag{5.5}$$

Hence, using the same notation as in (4.20),

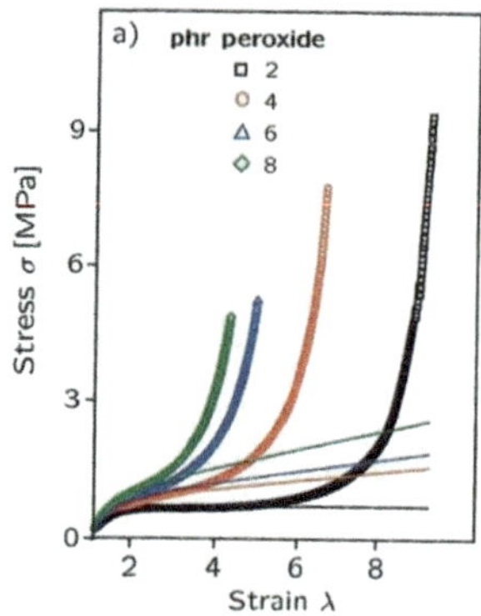

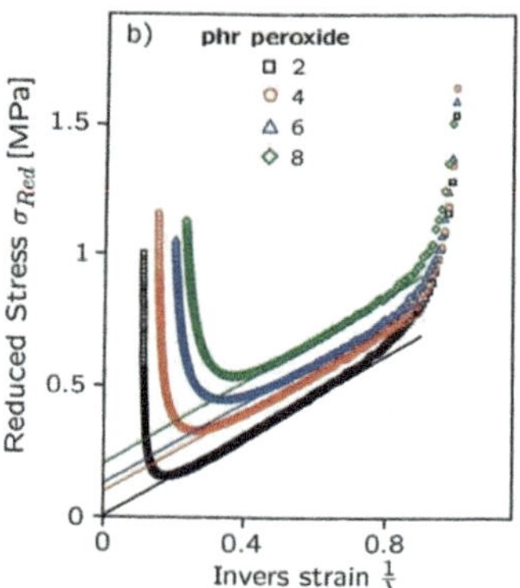

Fig. 5.3 Stress-strain measurements in standard form (**a**) and as Mooney–Rivlin plot (**b**). The material is a rubber (HNBR with 34% ACN) with variable concentration of a chemical cross-linking agent. The lines are fits based on (5.6) (**a**) and (5.7) (**b**), respectively. Reprinted with permission from [7]

$$\sigma_{zz} = \frac{1}{V}\frac{\partial \Delta F^{el}}{\partial \lambda} = 2\left(C_1 + \frac{C_2}{\lambda}\right)\left(\lambda - \frac{1}{\lambda^2}\right). \tag{5.6}$$

It is customary to define a reduced stress via

$$\sigma_{zz,red} \equiv \frac{\sigma_{zz}}{\lambda - \frac{1}{\lambda^2}} = 2\left(C_1 + \frac{C_2}{\lambda}\right). \tag{5.7}$$

Figure 5.3 shows stress-strain measurements in standard form (a) and as **Mooney–Rivlin plot** (b) compared to (5.6) (a) and (5.7) (b), respectively. The Mooney–Rivlin plot exhibits a nice straight portion over a significant range of values $1/\lambda$, which can be fitted with (5.7). The coefficient $2C_1$, which is the intercept with the reduced stress axis, is equated with Nk_BT and therefore yields the cross-link density. Here N is the number of chain segments joining (on average) two successive cross-links along a polymer chain. The number of cross-links is $N/2$ (think about a justification for this!).

The reasoning why $2C_1$ should be equal to Nk_BT is based on the comparison of our previous expression for the elastic entropy (3.53) to (5.5). Since (3.53) contains the factor $\lambda^2 + 2/\lambda - 3$ we take its coefficient (multiplied by $-T$) and equate it with the coefficient of the same λ-term in (5.5). This is hardly a 'clean' procedure however. For instance, note that the two expressions for the elastic free energy are different in the limit $\lambda \to 1$ (unless we ignore C_2 again) and neither expression is valid when λ becomes large. In other words, there is no easily identifiable limit in which $2C_1 = Nk_BT$ holds. Nevertheless, this is a commonly applied procedure.

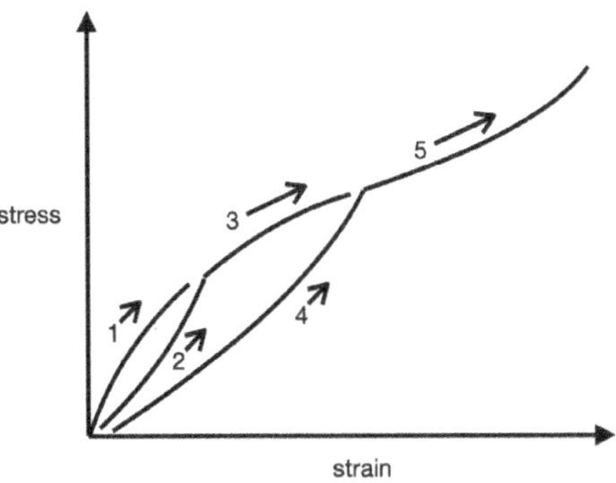

Fig. 5.4 Schematic illustration of the Mullins effect

Remark: The apparent deviation between the data and (5.7) in the limit $\lambda \to 1$ can be fixed by addition of an adjustable 'offset' $\sigma_{zz,o}$ to the right hand side of (5.6) (check this!).

● *Mullins Effect*

If a 'sufficiently elastic' polymer material is subjected to consecutive loadings, the stress, measured for a strain that was already reached in a previous loading, is found to be reduced. The cartoon in Fig. 5.4 illustrates this. The sample is strained and the stress follows path 1. Upon reaching the end of path 1, i.e. the largest strain during the initial loading, the sample is relaxed. During a second loading the stress does not increase along path 1, as before, but follows path 2 and then continues along path 3. When the end of path 3 is reached, the sample again is relaxed. During the third loading the stress increases along path 4 and then continues along path 5, etc. The curve corresponding to the combined paths 1, 3, 5,…is sometimes called the virgin curve. It is important to note that the strains used here are quite large (cf. the next figure). It is also important to note that this effect, called the **Mullins softening effect**, is most apparent in well vulcanized and/or filled samples (we discuss fillers in the next section). The next figure, Fig. 5.5, shows the result of an actual experiment. Here the sample experiences repeated stretching at constant strain amplitude before the latter is increased. Note that most of the softening is observed during the first deformation cycle. After a few additional cycles the rubber approaches a steady state or limiting cycle whose hysteresis is much reduced.

We do not want to discuss the Mullins effect in detail. A few comments must suffice: (i) the Mullins effect should not be confused with the so-called **Payne effect** occurring at much lower strain (cf. the next section); (ii) the generally accepted explanation for the Mullins effect is that high strain causes 'partially irreversible' (i.e. structural effects with very long relaxation times) movements of entanglements, network junctions or slippage of chain segments attached on filler particle surfaces as well as restructuring of filler particle networks and chain rupture; (iii) for more information the interested reader is referred to the aforementioned Ph.D. thesis by Plagge and to [9]. I also recommend the much older but insightful article *Network*

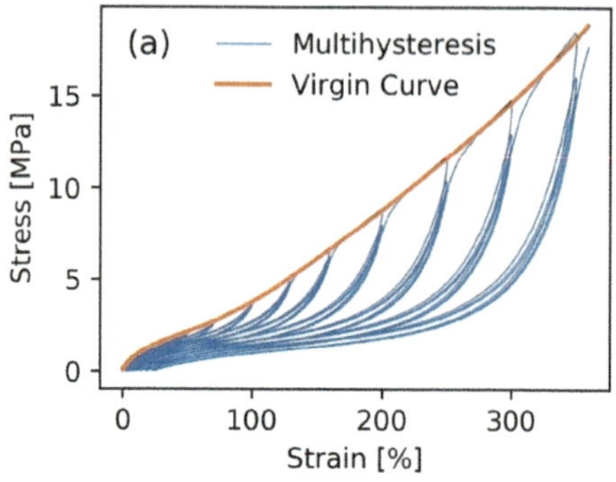

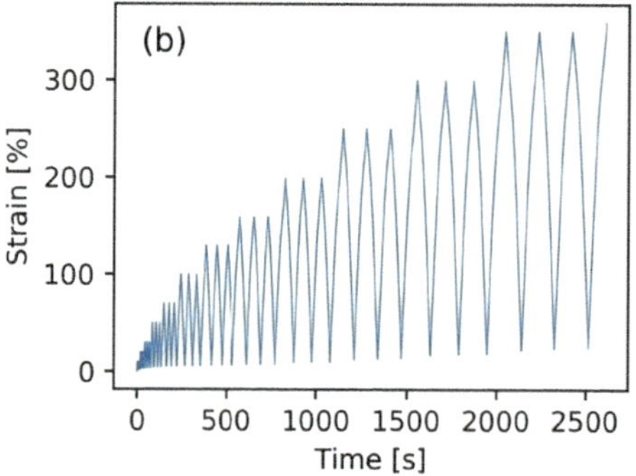

Fig. 5.5 **a** Example of a material experiencing the Mullins effect: 40 phr carbon black (N339) filled, sulfur cured EPDM. Continuous stretching (virgin curve) and multi-hysteresis experiment, consisting of repeated loading at increasing strain levels. **b** Corresponding strain protocol. Reprinted with permission from [8]

Theories of Reinforcement by Bueche [10].

Remark: You may wonder why the stress-strain curves depicted in Fig. 4.10 look so different from the ones depicted in Fig. 5.5. Figure 4.10 corresponds to an ideal linear model subject to a sinusoidal excitation. Figure 5.5 corresponds to a real material which is non-linear and the excitation has a saw-tooth shape rather than being sinusoidal. In addition, the hysteresis observed in stress-strain experiments has numerous contributions beyond the viscoelasticity of the neat polymer matrix. The material in Fig. 5.5 contains filler. We shall learn in the next section that filler networks can break and restructure when a load is applied. Moreover, load (and rate) dependent processes at the filler-polymer interface, e.g., chain slippage, contribute to hysteresis. When strain crystallization occurs, as we explain next, it also contributes to the observed hysteresis.

- *Strain-Induced Crystallization*

Strain-induced crystallization (SIC) is a process during which crystallites form within an originally amorphous polymer matrix in response to (usually large) strain (cf. the

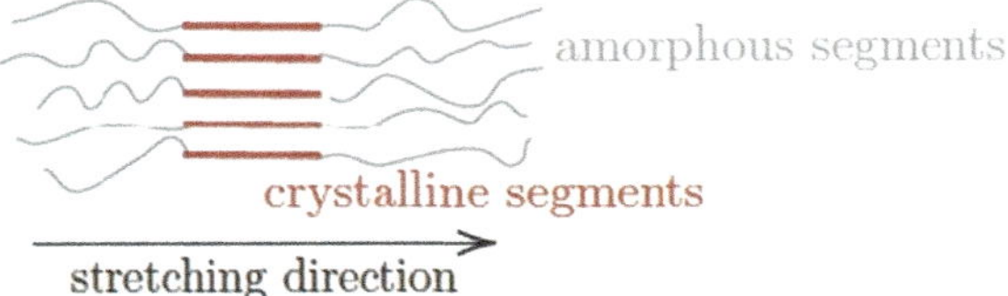

Fig. 5.6 Cartoon of a crystallite within an amorphous network. Reprinted with permission from [12]

cartoon in Fig. 5.6). SIC occurs in natural rubber (NR), as well as other elastomers and polymers—even though much of the interest in SIC is due to the importance of NR in technical applications (perhaps foremost among these are tires in particular for trucks).

Figure 5.7 shows stress-strain curves of NR together with the observed crystallinity during loading and unloading. To the superficial observer the curve in panel (a) looks somewhat similar to the stress-strain curves we have already discussed in

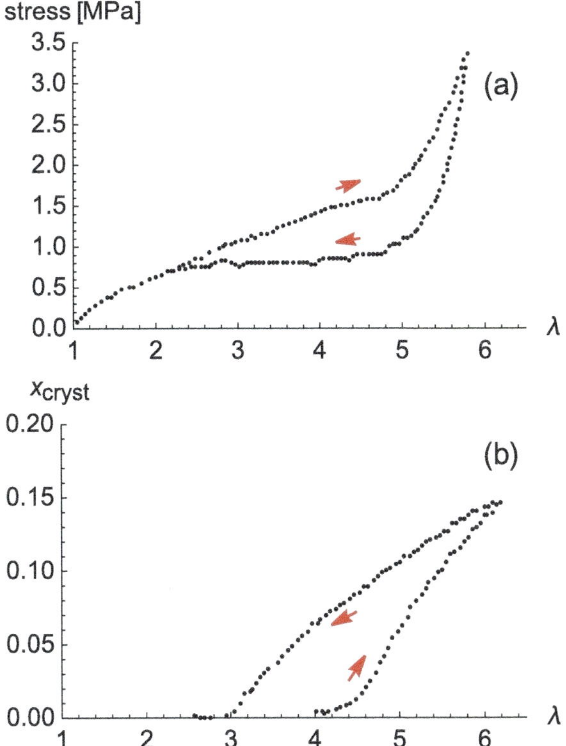

Fig. 5.7 a Stress-strain (λ) curves of natural rubber. The curves were obtained after the fourth mechanical cycle at a strain rate of 1 mm/min. **b** Crystallinity measured during the cycle in panel (a) at room temperature. Arrows distinguish the loading from the unloading part of the strain cycle. Adapted with permission from Fig. 2 in [13]. Copyright 2003 American Chemical Society

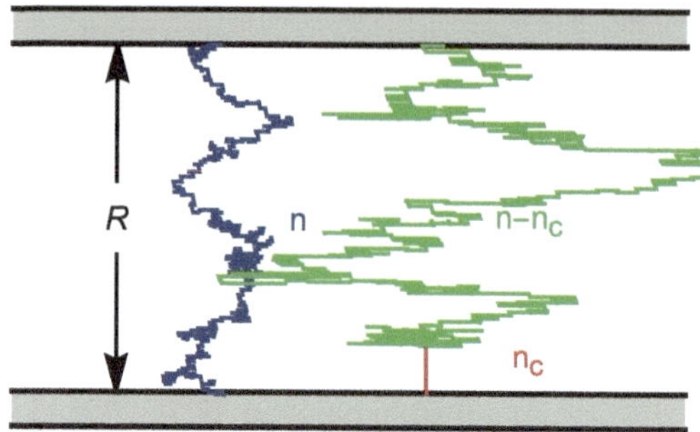

Fig. 5.8 Illustration of chain conformations without (blue) and with (green) strain crystallization (red)

the context of the Mullins effect. However, the curve depicted in Fig. 5.7 was obtained after a certain number of previous cycles and the central portion of in particular the unloading part of the cycle is quite flat. It looks very much like the transformation from one phase to another (e.g., gas-liquid or liquid-solid coexistence in the P-V plane). X-ray scattering in fact reveals the presence of crystallites (their size is roughly $10\,\mathrm{nm}$) formed beyond a certain strain threshold. Comparing panels (a) and (b) it appears that the stress is lower if the crystallinity is higher at any fixed strain.

Is there a theoretical argument why this should be the case? Yes—and, as one might expect, it was suggested by Flory [11].

However, before going into the details, it is useful to discuss the example in Fig. 5.8. The cartoon shows two ideal polymer chains containing an identical number of Kuhn segments n attached to opposite walls. The length of the Kuhn segments is b and the separation of the walls is R. It is assumed that $n\,b \gg R$. The entropic force on the anchoring points of the chain on the left is

$$f = -T\frac{dS(R)}{dR} \propto \frac{R}{n}, \tag{5.8}$$

where $S(R)$ is given in (2.68). The chain on the right, even though it possess the same number of Kuhn segments, differs from the first. It contains n_c consecutive Kuhn segments which are fully aligned and perpendicular to the walls. The rest of the chain, however, is a regular (amorphous) chain of the remaining $n - n_c$ Kuhn segments. At this point we are not interested in what causes the first n_c Kuhn segments to be special. Instead we want to know how these n_c Kuhn segments affect the force on the anchoring points of the chain.

Due to the constraint, the force f_c on the ends of the right chain is

$$f_c \propto \frac{R - n_c b}{n - n_c}. \tag{5.9}$$

Since the proportionality constants in (5.8) and (5.9) are the same, we must compare R/n to $(R - n_c b)/(n - n_c)$ to decide which force is larger. For simplicity we assume $b = 1$ and use the numerical example $R = 10$, $n = 100$ and $n_c = 4$. This yields

$$\frac{R}{n} = 0.1 \quad \text{and} \quad \frac{R - n_c b}{n - n_c} \approx 0.06, \tag{5.10}$$

Hence $|f_c| < |f|$. Now let's increase R from 10 to 12. If we want f_c to be constant, what does this imply for n_c? In order for f_c to remain constant, we must increase n_c from 4 to approximately 6. And this is essentially what we see in Fig. 5.7.

In other words, the formation of the crystallites, containing aligned polymer segments oriented predominantly in the direction of the applied strain/stress, 'buys' the remaining amorphous Kuhn segments conformational freedom, which they otherwise would not possess. This in turn reduces the force on the anchoring points of the chains in their network. SIC therefore is a means by which an elastomer material can reduce the stress acting on it. The crystallites can 'pop up' (on a time scale of ms) where needed and thus create a flexible response to large local strain. The crystallites melt when the strain is reduced. SIC therefore has important effects on strength and fatigue properties of the strain crystallizing material.

However, let's return to the aforementioned work of Flory. His is a single-chain theory described by a two-term free energy

$$\Delta F / T = -\Delta S^{el} + n(1 - x)\theta. \tag{5.11}$$

The first term is an elastic entropy given by

$$\Delta S^{el}/k_B = -\frac{1}{2}\left\{\left(\lambda^2 + \frac{2}{\lambda}\right)\frac{1}{x} - 3 - 2\lambda\frac{1 - x}{x}\sqrt{3n}\,p + \frac{(1 - x)^2}{x}3n\right\}. \tag{5.12}$$

Here the quantity x is defined via $n_c = n(1 - x)$. n_c is the number of straightened Kuhn segments in a chain composed of n Kuhn segments (cf. the above example). Note that for $n_c = 0$, i.e. $x = 1$, ΔS_{el} agrees with $\Delta S^{el}_{stretch}/N$ in (3.56). In Flory's paper $p = \sqrt{2/\pi}$. however, here we use a slightly adjusted value in order to achieve $n_c = 0$ for $\lambda = 1$ (see below).

But what is the meaning of the second term—in particular the quantity θ? It is due to straightening of $n_c = n(1 - x)$ consecutive Kuhn segments within a chain. If we force a 'crystallite' to exist above the melting temperature we must pay a positive 'penalty' in the form of work, i.e. by deforming the sample. The equilibrium x-value, i.e. x_o, follows via $\partial \Delta F / \partial x|_\lambda = 0$:

$$x_o^2 = \frac{\left(\frac{\lambda^2}{2} + \frac{1}{\lambda}\right)\frac{1}{n} + \frac{3}{2} - \lambda\sqrt{\frac{3}{n}}\,p}{\frac{3}{2} - \theta}. \tag{5.13}$$

If we require that $n_c = 0$, i.e. $x_o = 1$, for $\lambda = 1$ then

Fig. 5.9 **a** Crystallization $1 - x_o$, versus strain, λ. **b** Force, f, divided by temperature T, on a segment versus λ. Solid line: Equation (5.15); dashed line: $\lambda - 1/\lambda^2$. Parameter values: $n = 36$, $\theta = 0.19$ and p according to (5.14). This is Fig. 3 in [14]

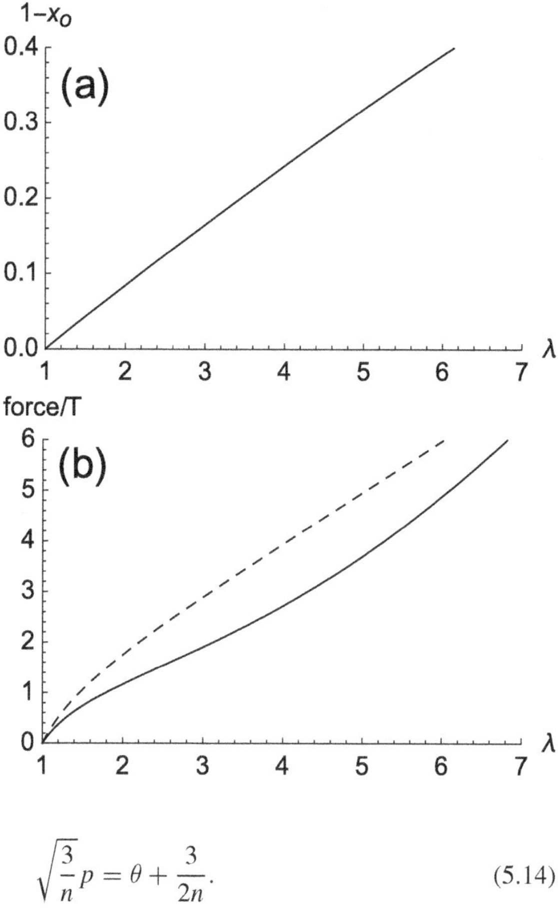

$$\sqrt{\frac{3}{n}}\,p = \theta + \frac{3}{2n}. \tag{5.14}$$

Finally, we find the average force f on a chain's ends via

$$f/T = -\frac{d\Delta F/T}{d\lambda} = -\frac{\partial \Delta F/T}{\partial \lambda}\bigg|_x - \underbrace{\frac{\partial \Delta F/T}{\partial x}\bigg|_\lambda}_{=0}\frac{dx}{d\lambda} \tag{5.15}$$

$$= \left(\lambda - \frac{1}{\lambda^2}\right)\frac{1}{x_o} - \sqrt{3n}\,p\left(\frac{1 - x_o}{x_o}\right).$$

Figure 5.9 shows an example where $n = 36$ and $\theta = 0.19$. Thus, from (5.14) it follows that $p \approx 0.803$ (very close to the aforementioned Flory value of 0.798). Like in the previous example, we find that the force at constant strain in a chain with SIC is less than that in a chain without SIC.

However, Fig. 5.9 is far from being a complete theory for experiments like the one illustrated in Fig. 5.7. Aside from the fact that the force curve in Fig. 5.9 does not show the pronounced upturn at high λ due to the absence of finite chain extensibility, there are other significant shortcomings. First, there is only a single curve and no hysteresis. This is because this is an equilibrium single-chain theory. Second, crystallization in this theory begins immediately. The experiments, on the other hand, find rather distinct and quite elevated strain values upon loading and unloading where crystallites are formed or disappear. This is because the time-dependent formation and cooperative behavior of crystallites is not addressed in Flory's theory. A suggestion for the rather distinct onset of crystallite formation is made in [14]. The authors propose that $1 - x_o$, shown in Fig. 5.9a, is the sum of two contributions, i.e. $1 - x_o = x_1 + x_{agg}$. Here x_{agg} is due to the crystallites observed via scattering experiments. x_1, on the other hand, corresponds to straightened chain segments in a largely amorphous environment, which are not yet part of a crystallite. Both x_1 and x_{agg} are described analogous to free and bound monomers in the context of reversibly assembly of labile polymeric aggregates—which in these lecture notes are discussed in Sect. 5.3. The theory of reversible assembly of molecular aggregates implies a critical concentration required for aggregates, or crystallites in the case at hand, to form. And this threshold concentration might just explain why there is such a distinct onset of strain induced crystallisation.

Subsequent to Flory's seminal work, numerous other researchers have made contributions to SIC. However, this is not the place for a comprehensive overview. Instead, the interested reader may want to consult [15–17] and the references therein.

- *Loss Tangent*

We had introduced the loss tangent or $\tan \delta$ in the context of (4.60), where it is defined as the ratio of the loss modulus to the storage modulus. Aside from the dynamic moduli themselves, the loss tangent is one of the quantities most likely measured during mechanical testing of a polymer sample. Since it is an important laboratory indicator for rubber performance in the tire industry, we want to briefly discuss it here. The importance of $\tan \delta$ as a tire rubber performance indicator is illustrated in Fig. 5.10.

The figure shows two experimental loss tangent curves plotted versus temperature. For the designer of tire tread materials the comparison between two such curves provides information about the relative performance of the materials in terms of rolling resistance, i.e. fuel efficiency, and/or breaking under wet and/or icy conditions. Just like the storage modulus and the loss modulus individually, the loss tangent is affected by almost all changes made to a rubber's recipe and processing. Being the ratio of two moduli, $\tan \delta$ is a dimensionless quantity. When filler, which we discuss in the next section, is added, the peak (usually) does not shift but the height of the peak is reduced (e.g., Figs. 2.63 and 2.64 in [7]). More or less the same happens when the amount of filler does not change but the filler particle size is reduced (e.g., [18]).

Figure 5.11 shows a picture of two rubber balls. If you just hold them in your hands or squeeze them, you can't really tell them apart. Their weight and hardness appears to be the same. But dropping them onto a hard surface reveals a pronounced

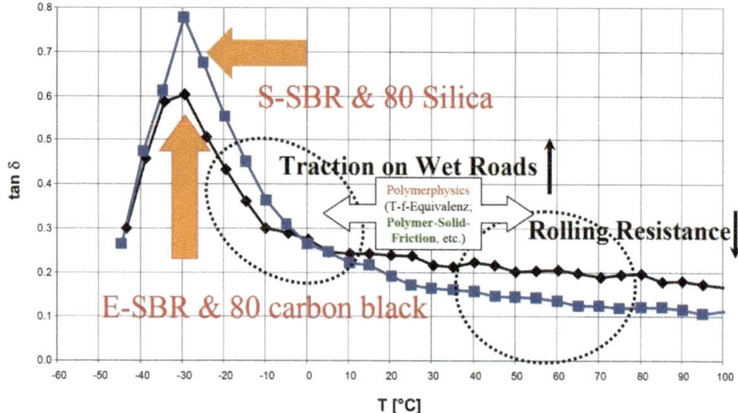

Fig. 5.10 Loss tangent versus temperature for styrene-butadiene rubber (SBR) filled with carbon black and silica nanoparticles. The strain amplitude is 10% and the frequency is 10 Hz. The figure is reprinted with permission from a talk given by G. Heinrich

Fig. 5.11 A bouncy and a not so bouncy rubber ball

difference. One of the balls bounces, whereas the other one hits the surface and does not bounce back into the air. I have been told, that candidates for a position in a well known tire company are regularly asked to explain this table top experiment during their job interviews. What is your explanation based on what you have learned thus far?

5.2 Filler Effects

Modern polymer based materials contain numerous components which are not polymers. Prominent among these components are particulate fillers. In the following discussion of why fillers are added and how they affect the behavior of the polymer matrix, we shall concentrate on elastomer materials used in the tire industry [19]. However, much of what is said applies to other polymer materials as well.

Figure 5.12 depicts a typical filler used in tire rubbers—carbon black. The top panel in the figure shows transmission electron micrographs or TEM images. Each TEM is based on a thin slice -usually 60–100 nm thick on average—cut from a material sample. What we see are aggregates of carbon black nanoparticles embedded in a polymer matrix. The aggregate's contrast varies, because in certain places there are particles stacked along the line of sight. But what is a particle? Usually the smallest distinguishable 'spheres' are the so-called **primary particles**. However, primary particles do not occur isolated from one another. Instead they are fused into small aggregates—let's call the **minimum aggregates**. These minimum aggregates usually do not break during the mixing process when the filler is introduced into the polymer matrix. The size of the minimum aggregates and the size of the primary particles appear to be correlated, i.e. their linear dimensions differ by roughly a factor of three [20]. Driven by the interface free energy, fillers show the tendency to form larger aggregates over time in a process called **flocculation**. Attractive dispersion forces between the filler particles let them form larger aggregates and structures beyond. The larger size and the weaker bonding means that the aggregates formed during flocculation can be broken when the material is mechanically deformed.

Fillers are characterized in terms of certain grades. In the case of carbon black different grades are distinguished by a code starting with the letters N or S followed by a number (see American Society for Testing and Materials ASTM D1765-21). The letters give indication of the influence of the carbon black on the rate of cure (vulcanization speed) of a typical rubber compound. The first digit is indicative of the (primary) particle size (or actually its surface area). Following digits contain additional information regarding the particle morphology. When the concentration of filler is increased, it will form a **spanning filler network** in the polymer matrix as shown in the middle panel of Fig. 5.12. Note that the polymer matrix itself usually is not conductive but carbon black is. The measurement of the filled polymer's **resistivity** versus carbon black concentration reveals a distinct step or filler **percolation threshold**. An illustration of a resistivity measurement is shown in the bottom panel of Fig. 5.12 (the real experiment is discussed in [21]). Note that the position of the percolation threshold depends on the system (cf. below). Here the unit of concentration is **phr** (parts per hundred) (in weight) rubber. The filler concentration in commercial elastomer materials varies but usually exceeds the percolation threshold.

Remark: Another experimental method which yields information about the filler structure inside a polymer matrix, in addition to TEM images, is small angle X-ray scattering. We discuss an example in Problem 5.2.

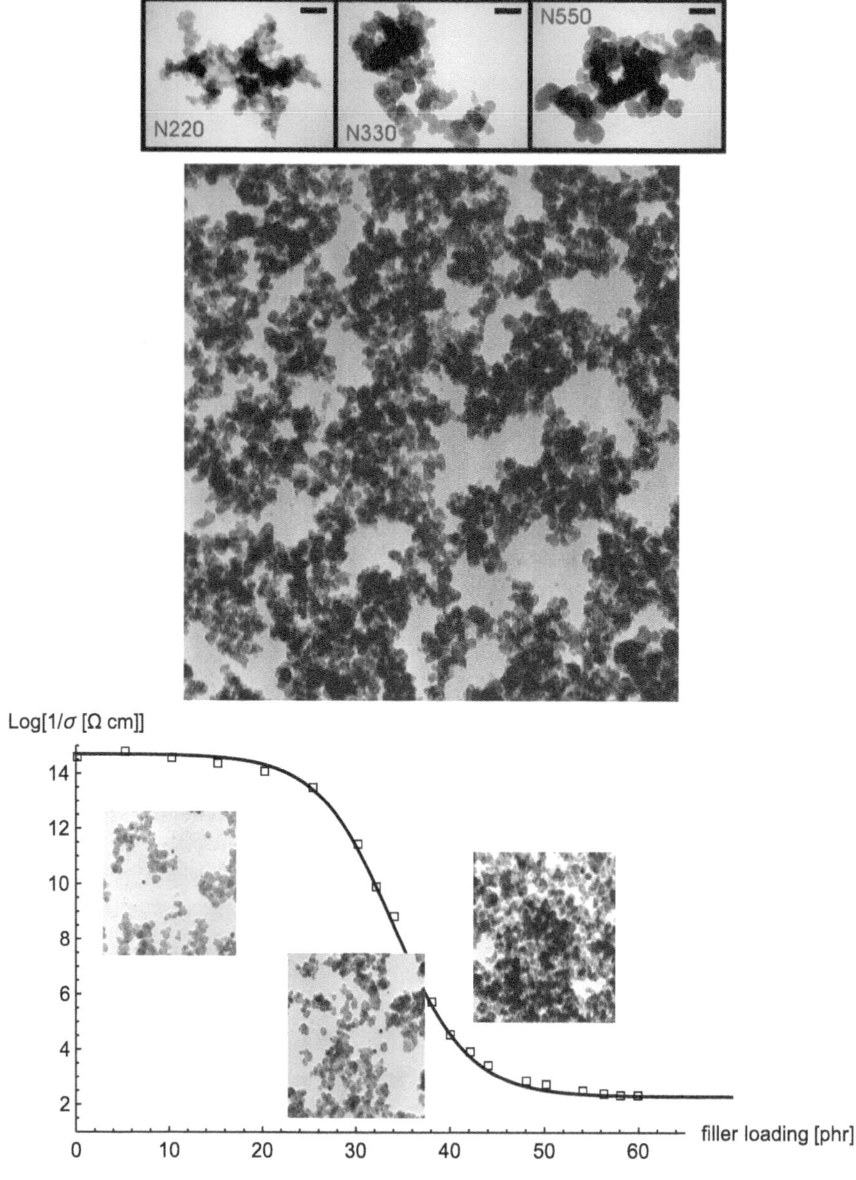

Figure axis labels:

Fig. 5.12 Top: Transmission electron micrographs (TEM) of three different grades of carbon blacks—N220, N330, and N550—disperse in a polymer matrix (bar length: 100 nm). Reproduced with permission from [22]. Middle: TEM-micrograph of a carbon black (N330) network obtained from an ultra-thin cut of a filled rubber sample. Reproduced with permission from [22]. Bottom: Illustration of the volume resistivity $1/\sigma$ of a filled elastomer versus filler concentration in phr. The inserts represent TEM images at the respective filler concentration along the x-axis

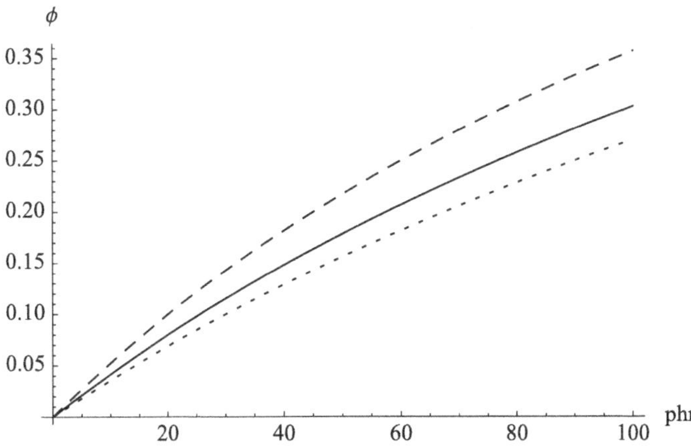

Fig. 5.13 Filler loading: conversion between phr and volume fraction ϕ. Dashed line: carbon black; solid line: silica; dotted line aluminum ($c_{Al} = 2.7\,\mathrm{g\,cm^{-3}}$)

In the following we shall use the volume fraction ϕ to measure the fractional amount of volume occupied by filler. Hence, we shall need to convert between phr and ϕ. As we have already stated, phr is the abbreviation of *parts per one hundred base polymer*, i.e. the unit phr means proportion by weight, with respect to 100 parts of the crude rubber used. Carbon black has a density of around 1.8–2.1 $\mathrm{g\,cm^{-3}}$ and silica, another prominent filler (cf. below), has a density of around 2.2–2.3 $\mathrm{g\,cm^{-3}}$, whereas rubber densities are close to 1 $\mathrm{g\,cm^{-3}}$. The conversion from phr to filler volume fraction, ϕ, is

$$\phi = \frac{\dfrac{phr}{c_{filler}}}{\dfrac{phr}{c_{filler}} + \dfrac{100}{c_{rubber}}} \tag{5.16}$$

(neglecting additional components). Here c_{filler} and c_{rubber} denote the filler and the rubber density, respectively. The formula is shown in Fig. 5.13 for $c_{filler} = 2.7\,\mathrm{g\,cm^{-3}}$ (dotted line), $c_{filler} = 2.3\,\mathrm{g\,cm^{-3}}$ (solid line), and $c_{filler} = 1.8\,\mathrm{g\,cm^{-3}}$ (dashed line) using $c_{rubber} = 1\,\mathrm{g\,cm^{-3}}$.

Carbon black nanoparticles are not the only filler used. The second prominent filler system is the silica-silane system. Silica particles consist of amorphous SiO_2. The surface of silica nanoparticles is covered with silanol groups, i.e. various types of OH-groups. In order to improve the dispersion of silica in most polymers, the surface of the silica particles is compatibilized using silanes. The silane molecules are chemically bonded to the silica surface and usually also to the polymer matrix. Hence the silica-silane system is quite versatile and can be modified to optimize material performance. Aside from carbon black and silica there are numerous other fillers including nanotubes, organic particles, etc.

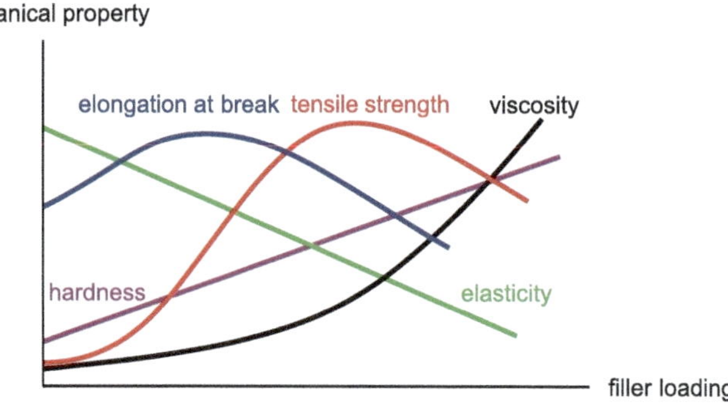

Fig. 5.14 Mechanical properties of elastomers versus filler loading

Figure 5.14 compiles the effect of variable filler concentration on a number of mechanical quantities. It is the goal of the theory of course to provide explanations for these observations.

Suppose we use the rheometer depicted in Fig. 4.9 to carry out a shear experiment on one of the polymers whose storage modulus master curve is depicted in Figs. 4.12 and 4.13. Let's assume the shear frequency is 1 Hz, which means we are on the plateau in Fig. 4.12b or, if the molecular weight is sufficiently large, in Fig. 4.13 as well. In the following G_o is this plateau value of the storage modulus. If the polymer matrix also contains filler G_o is replaced by G.

What happens if we keep the shear frequency constant at 1 Hz but change the shear amplitude? The qualitative answer is depicted in the top panel of Fig. 5.15. Without filler essentially nothing happens. In the case of a linearly elastic network the storage modulus does not depend on the amplitude of deformation. However, with the addition of filler this changes. An increase of the storage modulus is observed, which has several contributions. First there is a **hydrodynamic effect**. Then there is a **in-rubber-structure effect**. These two effects depend weakly on the deformation amplitude. However, the third contribution, denoted here as 'filler-filler interaction', strongly depends on the deformation amplitude. In the limit of small deformations and for filler concentrations near or above the percolation threshold it is by far the largest of all four of the shown contributions.

The reason behind the strongly non-linear behavior of the storage modulus is explained in the bottom panel of Fig. 5.15. The explanation starts with the spanning filler network still intact at small strain (roughly <1% strain). Somewhere between 1 and 10% strain the network is broken up into smaller pieces. But why should the network begin to break up already at 1% strain—and sometimes even at strains as low as 0.1%?

Before we address the question directly, let's construct a (local) theoretical picture of the filler network. The left panel in Fig. 5.16 depicts filler aggregates (shown as disks) forming network branches. The branches are embedded in tightly bound polymer sheaths. Filler branches plus their polymer sheaths together constitute **filler**

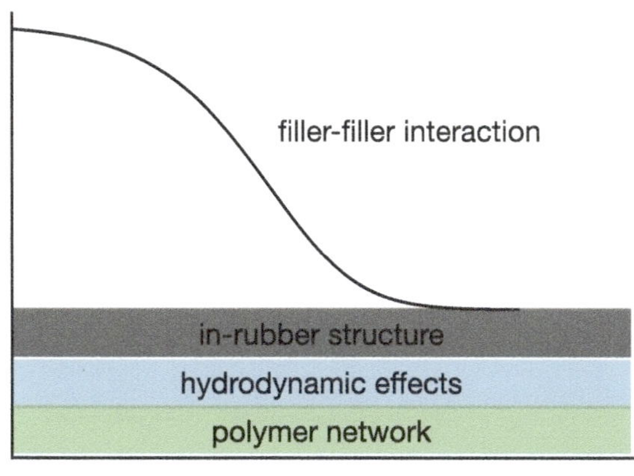

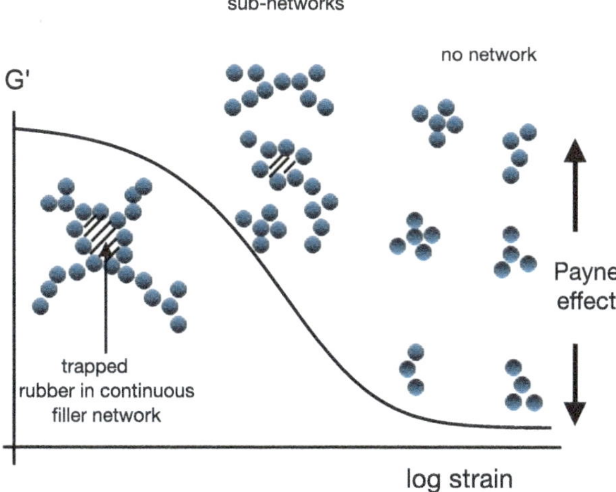

Fig. 5.15 Top: Contributions to the storage modulus. Bottom: Breakdown of the filler network with increasing strain. Adopted from [23]

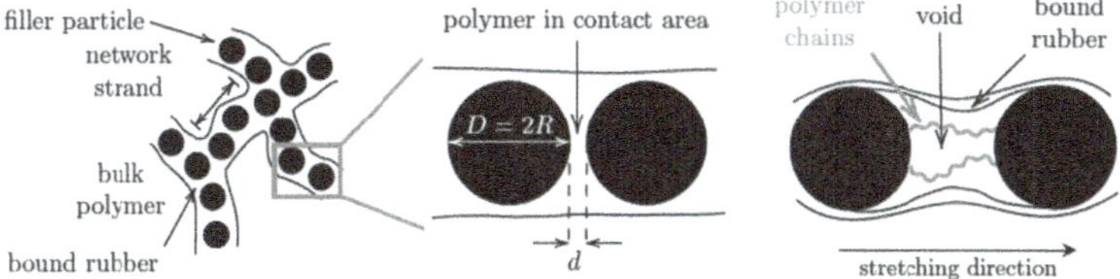

Fig. 5.16 Left: Filler strands within bulk-like polymer; middle: a filler-filler contact; right: voids between aggregates caused by straining the network strands. Reprinted with permission from [12]

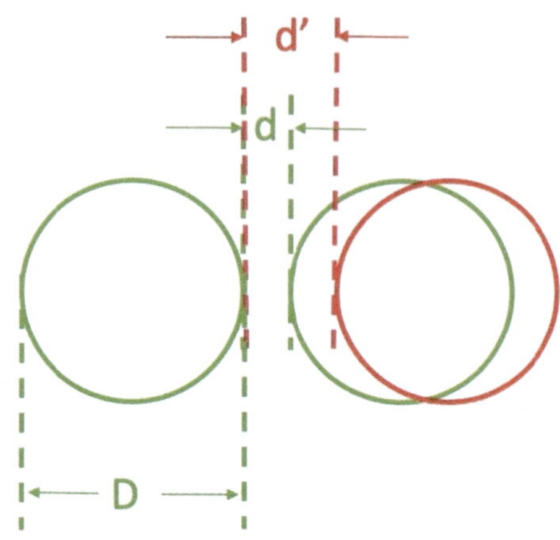

Fig. 5.17 Cartoon illustrating strain amplification in filler-filler contacts

network strands. The polymer beyond is bulk-like. A central element is the **contact**, depicted in the middle, consisting of adjacent filler particles plus surrounding amorphous 'sheath polymer'. When the contact is stretched, voids between the aggregates will occur at some point.

Note that the close proximity of neighboring filler particles causes pronounced **strain amplification** as illustrated in the cartoon depicted in Fig. 5.17. The green circles are two neighboring filler particles, whose diameter is D, in the unstrained sample. The resistivity experiments suggest that the gaps d between many of the filler particles must indeed be very small (<1 nm), i.e. $d \ll D$. Since the filler particles are not deformed when the sample experiences an external force, it must be the gap that widens from d to d'. Therefore the strain γ on a (one-dimensional) volume element containing the filler particles is related to the strain inside the gap γ_{gap} via

$$\gamma = \frac{d' + D - (d + D)}{d + D} \approx \frac{d' - d}{D} = \frac{d}{D}\gamma_{gap}. \tag{5.17}$$

Hence, $\gamma_{gap} \gg \gamma$. This calculation pertains to the cartoon. Nevertheless, the general idea applies also in a real network.

In the aforementioned strain range, the storage modulus undergoes a marked reduction commonly denoted as **Payne effect**. Somewhere between 10 and 100% strain the storage modulus levels off. The filler-filler interaction and the in rubber structure are mostly gone. Nevertheless, it is important to note that whatever is left is still significant, because the tensile strength of a filled polymer network considerably exceeds the tensile strength of the pure network (cf. Fig. 5.14) (see for instance [24]). The amorphous polymer sheath in Fig. 5.16 is replaced by a multitude of taut polymer chains anchored in part on the filler particle surfaces (depicted in Fig. 5.18).

In the following we want to discuss selected theoretical concepts, which will help us to better understand the mechanical behavior of filled polymer matrices. We shall talk about 'particles' a lot—meaning minimum aggregates.

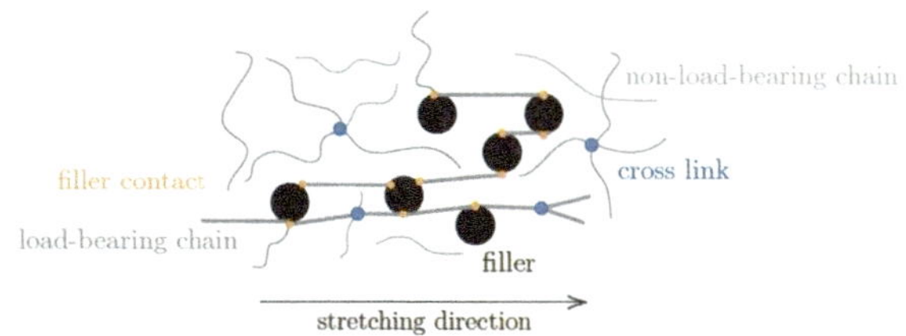

Fig. 5.18 Cartoon of load bearing polymer chains occurring at large strains. Reprinted with permission from [12]

The aforementioned hydrodynamic effect primarily refers to the continuum approach initially introduced by Einstein in a different but related context [25]. Einstein devised a method for determining molecular dimension via the viscosity increase when 'large particles' are added to a fluid of 'small particles'. Subsequently Smallwood [26] published a formally identical equation:

$$G = G_o \left(1 + \frac{2}{5}\phi \right).$$ (5.18)

Here ϕ is the filler volume fraction (The factor 2/5 is missing in Einstein's original calculation. He later notices and corrects the mistake.). This result is valid in the limit of small ϕ, assuming that the filler particles are hard spheres and employing 'no-slip' boundaries on their surface. It is obtained by solving the elastic equilibrium condition of the rubber treated as isotropic elastic material with filler particle inclusions. Generalizations of this equation to a power law series in terms of ϕ are discussed by Guth [27]; in particular we mention the expression due to Guth and Gold [28]:

$$G = G_o \left(1 + \frac{2}{5}\phi + 14.1\phi^2 \right).$$ (5.19)

While $G = G_o(1 + (5/2)\phi$ is an exact result in the above limit, which is an appealing feature, it has major shortcomings. Perhaps most seriously, (5.18) and (5.19) predict that the compound modulus is independent of the size of the filler particles. This is incorrect for modern sub-micron fillers. In the following we shall discuss the filler induced increase of the shear modulus from another angle.

In contrast to the no-slip boundary conditions underlying the Smallwood equation, we assume that the adhesion of the matrix material on the filler surface can be thought of as being due to surface bonds contributing strength analogous to the elastomer's bulk cross-links. But what do we mean by **'bonds'**? An example is a polymer segment physisorbed on the surface of a filler particle. The same polymer may possess another segment on a neighboring particle and thus connects the two particles. Or the polymer may be cross-linked to another polymer connected to the neighboring particle, etc. Another example is the covalent bonding between a particle's surface and the polymer matrix via a silane molecule. Yet another example is a hydrogen bond formed between the silanol groups of two neighboring silica particles. Hence there are many types of such molecular bonds. Of course, there are also dispersion forces between neighboring particles involving all atoms in the two particles (cf. for instance [29]). However, in most cases the latter interactions are not the dominant interactions affecting the mechanical properties of the material. It is important to note that the molecular bonds have their own temperature dependence (see for instance [30]). This means that a (highly) filled polymer matrix obeys time-temperature superposition to a lesser extend than an unfilled one.

Let us model the filler as composed of isolated monodisperse spherical particles with radius R. Under these conditions the modulus increase, in the limit of small shear amplitude, due to the particle-matrix interface should scale as ϕR^{-1}, i.e.

$$G - G_o \sim \frac{\phi}{R}. \tag{5.20}$$

Again we find a proportionality to ϕ. But we find also that a small particle size increases G. The derivation is as follows: N_P is the number of particles with volume v_P and surface area, a_P. Thus we have $N_P v_P = \phi$. Likewise we have for the total surface area A of the particles $N_P a_P = A$. With $v_P \propto R^3$ and $a_P \propto R^2$ one finds

$$A \propto \frac{\phi}{R}. \tag{5.21}$$

Because the increase of G_o should be due to the surface bonds, we arrive at the above conclusion. Figure 5.19 shows the storage modulus of carbon black filled polybutadiene in the limit of small strain plotted versus the inverse diameter of the minimum aggregates. These data follow the predicted $1/R$-dependence quite nicely. However, the volume fraction filler in this experiment is 0.18. This is close to the percolation threshold and the assumption of isolated aggregates is not valid. Unfortunately, this author was not able to find data for G' versus $1/R$ at filler volume fractions significantly below the percolation threshold.

However, there is more information which can be gained by our simple reasoning. We can argue that there is an upper limit beyond which particle size has an adverse effect on reinforcement. This is because the volume of a large particle can displace more bulk cross-links in the elastomer compared to the increase of the number of surface bonds. This is illustrated by the sketch in Fig. 5.20. Note that the number of bulk cross-links is proportional to R^3, whereas the number of surface bonds is proportional to R^2. We may estimate the cross-over radius, R_c, based on rough estimates of both the surface bond density, n_s, and the bulk cross-link density, expressed in cross-links per rubber monomer, ν. With $1/50\,\text{Å}^{-2} < n_s < 1/20\,\text{Å}^{-2}$ (in the case of silica) and $1/500 < \nu < 1/200$ we find $0.1\,\mu\text{m} < R_c < 1\,\mu\text{m}$, which is in accord with experimental observation (cf. page 11 in [31]). Small filler particles far below this limit are called **active**.

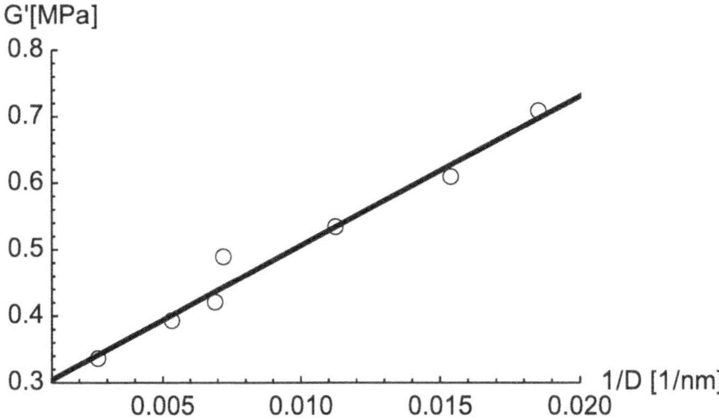

Fig. 5.19 G' at small strain amplitude versus the inverse diameter $1/D$ of the minimum aggregates according to Fig. 8 and Table 1 in [18]. Shown here is G' for polybutadiene at a constant volume fraction of 0.18. The Measurements were performed at $T = -70\,°\mathrm{C}$ and $\omega = 31.4$ rad/s

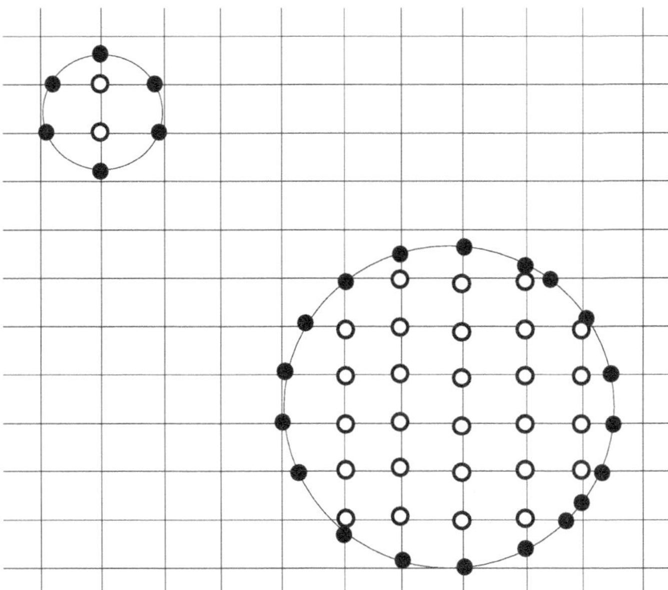

Fig. 5.20 Open circles: bulk cross-links; solid circles: surface bonds. Reprinted with permission from [24]

Next we want to study the effect of the filler structure near the percolation threshold. Generally, the shear modulus, in the limit of small strain, does not increase linearly. Instead a power law is observed, i.e.

$$G - G_o \sim \phi^y. \tag{5.22}$$

The exponent y can be significantly larger than unity. Let us construct an explanation for (5.22) and find y.

We consider N filler particles in a volume $V = L^3$. Hence

$$\phi \propto \frac{N}{L^3}, \tag{5.23}$$

where ϕ is the filler volume fraction. In a uniform system we expect in addition $N \propto L^3$, i.e. the filler volume fraction does not depend on L. This means that ϕ is an intensive quantity. However, if we assume instead

$$N \sim L^{d_f} \tag{5.24}$$

with $d_f < 3$, then the situation is different. The system now contains 'filler voids' on all length scales. Such cluster structures are called **fractal** structures and d_f is their mass **fractal dimension**. By combination of (5.23) with (5.24) we find

$$\phi \sim L^{-(3-d_f)} \to 0 \quad \text{for} \quad L \to \infty. \tag{5.25}$$

This means that a fractal filler distribution on all length scales throughout the rubber is not realistic!

Mechanical mixing results in a uniform system on a macroscopic scale, i.e. the mixing process does not allow a fractal filler network beyond a certain linear dimension. In this case ϕ is finite of course. Because this is an important aspect let us be more specific and call this length ξ, i.e. our material is a filled polymer matrix containing fractals of size ξ, which in turn contain N_ξ filler particles. Hence

$$\frac{N}{N_\xi} \propto \frac{L^3}{\xi^3} \tag{5.26}$$

and therefore

$$\phi \propto \frac{N_\xi}{\xi^3}. \tag{5.27}$$

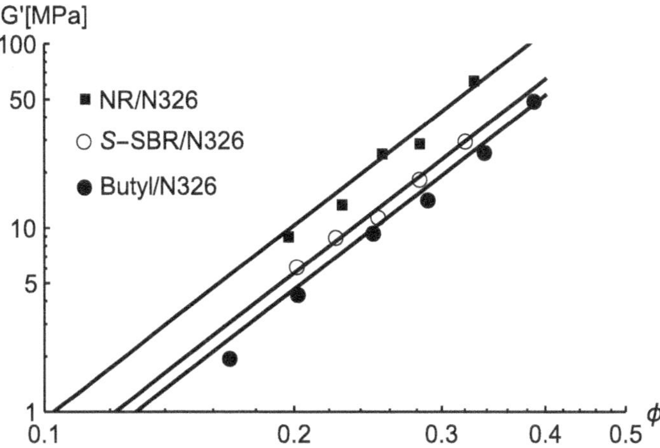

Fig. 5.21 Small strain storage modulus versus filler volume fraction for a variety of carbon black filled composites. The solid lines possess slope 3.5. The data are from Fig. 39 in [22]

Consequently we arrive at

$$\frac{1}{A_\xi} \sim \phi^{\frac{2}{3-d_f}}, \tag{5.28}$$

where $A_\xi = \xi^2$.

What is the significance of the expression (5.28)? The quantity A_ξ is the cross-sectional area of a fractal cluster containing N_ξ particles. Thus we may consider $1/A_\xi$ as an approximate measure of the density of filler-filler contacts in a (shear) plane cross-section through a fractal patch. Because there are on average L^2/A_ξ fractal patches in the entire sample cross-section, we may consider the product of L^2/A_ξ with $1/A_\xi$ as a measure for the shear strength of the entire sample cross-section. In order to obtain the modulus we divide by L^2, which yields

$$G - G_o \sim \frac{1}{A_\xi}\frac{L^2}{A_\xi}\frac{1}{L^2} \sim \phi^{\frac{4}{3-d_f}}. \tag{5.29}$$

Of course, this should hold only if ϕ is large enough for the spanning filler network to form.

Meakin et al. [32] have studied the collisions between point masses and fractal aggregates via computer simulation of different models. They obtain d_f-values ranging from 1.8 to 2.1 depending on the model (i.e. $3.3 < y = 4/(3 - d_f) < 4.5$). The resulting structures all look very similar to experimental filler structures obtained from transmission electron micrographs. Thus, a certain variation of the exponent y appears permissible. Figure 5.21 shows an experimental confirmation, where $y \approx 3.5$, of the picture developed here.

The fractal concept we have just outlined is not new. Significant insight was originally provided via computer simulation. For instance, the cluster-cluster aggregation model introduced by Meakin [33] and Kolb et al. [34] initiated an upsurge of interest in the properties of aggregated colloids. Subsequent simulation studies suggested that colloidal aggregates behave as stochastic mass-fractals on a scale large compared to the primary particle size. Experimental studies did confirm this and provided values for the fractal dimension d_f in good agreement with those obtained via simulation [35–38].

Our final expression (5.29) is close to one proposed by Brown [39], i.e. $y = (3 + d_{chem})/(3 - d_f)$, where d_{chem} is the so-called chemical length exponent. An in depth discussion of the development and application of fractal networks in the context of colloidal systems can be found in [40]. Full appreciation of these ideas in the context of filled rubbers came with the works of Heinrich, Klüppel, and Vilgis [41, 42] (and references therein).

Strictly speaking the above applies close to the percolation threshold only. At filler volume fractions far above the percolation threshold we must expect a different and likely less fractal structure. What would by the result if the filler was forming completely random connections? We consider N filler particles distributed over L^3 cells in a cubic grid. The probability that a particular particle finds a particle in a neighboring cell is (neglecting correlations) qN/L^3, where q is the coordination number of the lattice. Because the number of pairs formed in this fashion is proportional to N^2/L^3 and the density of pairs is proportional to $N^2/L^6 = \phi^2$, we expect

$$G - G_o \sim \phi^2 \tag{5.30}$$

in the case of a random filler distribution. Indeed, many measurements of $G - G_o$ versus ϕ yield y-values close to 2.

We want to return to the dependence on particle size in the context of a dense filler network. Let's assume we compare systems composed of spherical particles whose radii are R and $2R$, respectively. In both systems the filler network structure is the same and the filler volume fraction is also the same. This means that we can transform the system containing the smaller particles into the system containing the larger particles by scaling its axes by a factor of 2. Within a fixed area in the shear plane, i.e. this area is not scaled, this reduces the number of particle-particle contacts by a factor of 4. Our premise is that $G - G_o$ is proportional to the number of particle-particle contacts in a shear plane and thus $G - G_o \propto R^{-2}$. However, the interaction between particles in a contact also depends on R. This is illustrated in the sketch shown in Fig. 5.22. If the forces bonding the spheres act near the surface only (e.g., short chain segment physisorption or hydrogen bonds), i.e. we assume a certain range between d and $d + 2h$, then the interacting surface area S is given by

$$S = 2\pi Rh. \tag{5.31}$$

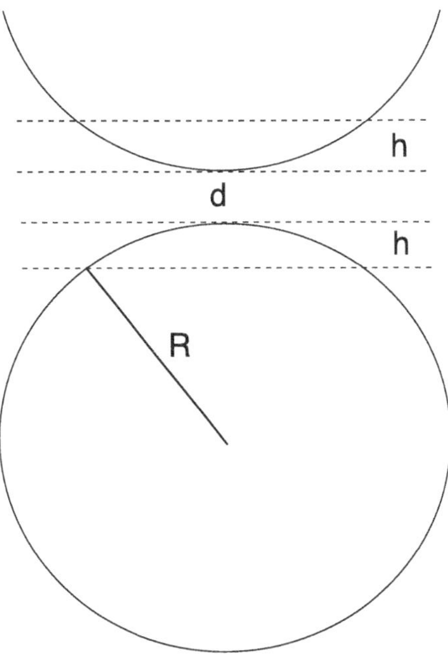

Fig. 5.22 Primary filler particles pictured as spheres at close proximity

But even if we consider the interaction to be contributed from atomic pair-interactions throughout the spheres, the resulting interaction still is proportional to R [29] (the force between two spheres of equal radius R is $-\frac{A}{12}\frac{R}{d^2}$, where A is the **Hamaker constant** and d ($d \ll R$) is a small surface-to-surface separation between the spheres). Therefore all in all the two contributions yield a R^{-1}-dependence, i.e.

$$G - G_o \sim \frac{\phi^y}{R}. \tag{5.32}$$

This R dependence is the same which we had found previously in the limit of isolated particles—and this time Fig. 5.19 can be viewed as a confirmation.

The position of the percolation threshold along the volume fraction axis, ϕ_c, does depend on the particle grade / size. An example is shown in Fig. 5.23. How can we understand this dependence? In (5.28) the length ξ is expressed in units of particle size R, i.e.

$$\frac{R^2}{(\xi R)^2} \sim \phi^{\frac{2}{3-d_f}}. \tag{5.33}$$

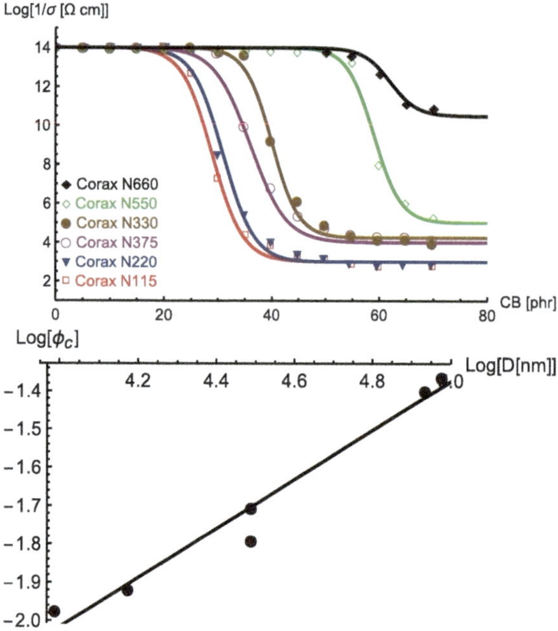

Fig. 5.23 Top: Resistivity data across the percolation threshold for different grades of carbon black from Carbon Blacks for Electrically Conductive Rubber Products, Technical Report Degussa AG (TR812); CD *Solutions for the Rubber Industry* (Remark: the company is now Evonik and the (nice) technical reports are difficult to obtain, unfortunately.). The polymer is SBR 1500. CORAX is a trademark of Orion. The lines are based on a suitable fit function, which itself has no particular physical meaning. Bottom: determination of the exponent $3 - d_f$ from a plot of $\log \phi_c$, where ϕ_c is the filler volume fraction at the inflection point of the resistivity data, versus $\log D$, where D is the corresponding aggregate size taken again from Table 1 in [18]

We have stated that the length scale over which fractal filler structure can exist is determined by ξ or rather by ξR in absolute units. It is plausible to assume that this length scale is similar for different R, i.e. $\xi R \approx const$. Hence, we find

$$\phi_c \sim R^{3-d_f}. \tag{5.34}$$

This means that the percolation threshold shifts to larger filler volume fractions when the filler particles are larger—in agreement with the experimental observation in Fig. 5.23. The bottom panel in Fig. 5.23 is a test of this relation. The slope of the linear fit to the data is about 0.6 and thus $d_f \approx 2.4$. This is larger than what we have discussed before and larger than the d_f-value needed to obtain $y \approx 3.5$. However, $y \approx 3.5$ is far from universal and larger values are found as well (cf. Fig. 8 in [43]). In addition, the conversion from phr to volume fraction is approximate and so are the particle sizes.

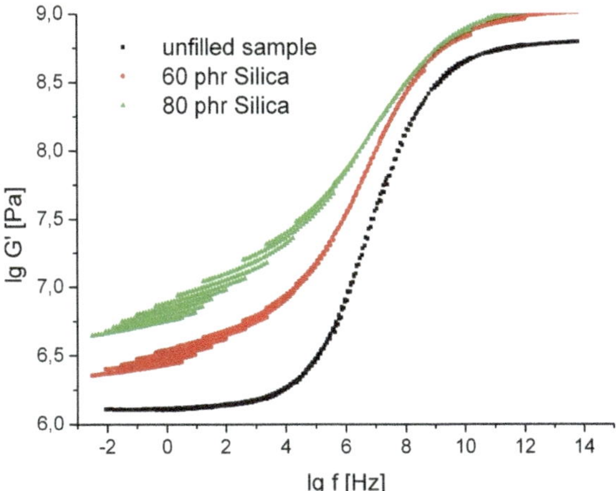

Fig. 5.24 Master curves of the storage modulus G' of an unfilled S-SBR sample and the same samples filled with 60 and 80 phr of silica created with (horizontal) shift factors from the unfilled sample. Reprinted with permission from [44]

Before leaving this section we want to mention the effect of fillers on time-temperature superposition. Figure 5.24 shows master curves of the storage modulus G' of an unfilled S-SBR sample and the same samples filled with 60 and 80 phr of silica created with (horizontal) shift factors from the unfilled sample. While (4.73) and (4.72) work very well for unfilled samples, as we have seen numerous times before, they do not work equally well when there is filler in the polymer. Why is this? The temperature dependence of our shift factor a_T rests on the assumed linear temperature dependence of the free volume. By adding filler we introduce interfaces into the system. These interfaces, between filler and polymer and between filler particles, contain bonds (cf. above). In the case of silica, for instance, we find hydrogen bridge bonds. These hydrogen bridge bonds may connect adjacent particle surfaces or they may connect an OH-group on a silica surface to a water molecule above the surface (in the case of silica, a certain amount of water, perhaps a couple of % by filler weight, is always present). In any case, the number of such bonds depends on temperature akin to the equilibrium constant in a chemical reaction (cf. [30]). Since the number of (reversible) bonds does affect the coupling of the filler particles among themselves as well as their coupling to the polymer matrix, it may not be surprising that this new temperature dependence, which is different from the temperature dependence of a_T in a pure polymer system, does interfere with time-temperature superposition as we had discussed it. Improved master curves in filled systems can be obtained if the horizontal shift factor a_T is supplemented by a second vertical shift factor (see [44]).

Finally, we have not yet and will not discuss filler particles possessing anisotropic shape, e.g., **carbon nanotubes** (CNTs). Particularly CNTs have been the focus of intensive research for some time. Even in small amounts (a few % by weight) they reinforce an elastomer matrix considerably. However, fillers possessing high shape anisotropy can be difficult to disperse properly. They tend to align and aggregate (we shall learn something about the underlying physical principles in the section on liquid crystallinity), which may not always be desired. In addition, the interface between the CNTs and the polymer matrix poses additional new challenges. The interested reader may look up [45].

Even though this discussion of filler effects is rather limited, it highlights that fillers do have a significant influence on the properties of a polymer material. It should also be clear that reliable or quantitative (and sometimes even qualitative) theories for the prediction of filler effects do not exist at this time.

5.3 Stable and Labile Liquid Crystalline Polymers

Stable Liquid Crystalline Polymers:

A summary of what we discuss next is compiled in Fig. 5.25. The cartoon at the top shows randomly oriented, persistent flexible polymers in solution at low concentration. Remember that in Sect. 2.1 we had mentioned polymers which can be described as homogeneously bend-elastic. One example, PBLG, is shown in Fig. 2.8. The quantity characterizing their flexibility is the persistence length P defined in (2.22). Upon increasing the concentration, the polymers are forced to align their contour tangent vectors $\vec{t}$ along a spontaneously chosen direction, which is called the **director**. Here we consider polymers with fairly short-ranged interactions. To good approximation they can be described as locally hard but overall flexible cylindrical rods. What forces the alignment is the excluded volume of the rods. Specifically, it is the competition of **orientation entropy** versus **packing entropy**. The alignment is measured using the orientation distribution function $f(\theta)$, where θ is the angle of $\vec{t}$ relative to the director. If the orientation is isotropic, then $f(\theta)$ is constant. Otherwise it exhibits maxima at $\theta = 0$ and $\theta = \pi$. These two types of orientational ordering correspond to two phases, i.e. the **isotropic phase** and the **nematic phase**, respectively. The description of the transition from the former to the latter is our main objective.

However, other so-called **liquid crystalline phases**, featuring both orientational as well as **translational ordering**, may occur as well. Two examples are shown at the bottom of Fig. 5.25. In a **hexagonal columnar phase** the centers of mass of the rods tend to form a hexagonal lattice if we look along the rods' axes. If we look at the rods from the side, their centers of mass appear randomly distributed. In this sense a columnar phase can be described as a combination of a two-dimensional solid with one-dimensional liquid. A second example of a liquid crystalline phase is a **smectic phase**. In a smectic phase the molecules form layers. Hence a smectic phase may be described as the combination of a two-dimensional liquid parallel to the layer boundaries with a one-dimensional crystal in the perpendicular direction. If concentration (via excluded volume) is the main driving variable taking the system from one phase to another we call this type of liquid crystalline behavior **lyotropic**. If temperature is the main driving variable we call that type of liquid crystalline behavior **thermotropic**. However, this distinction is not absolute and lyotropic liquid crystals

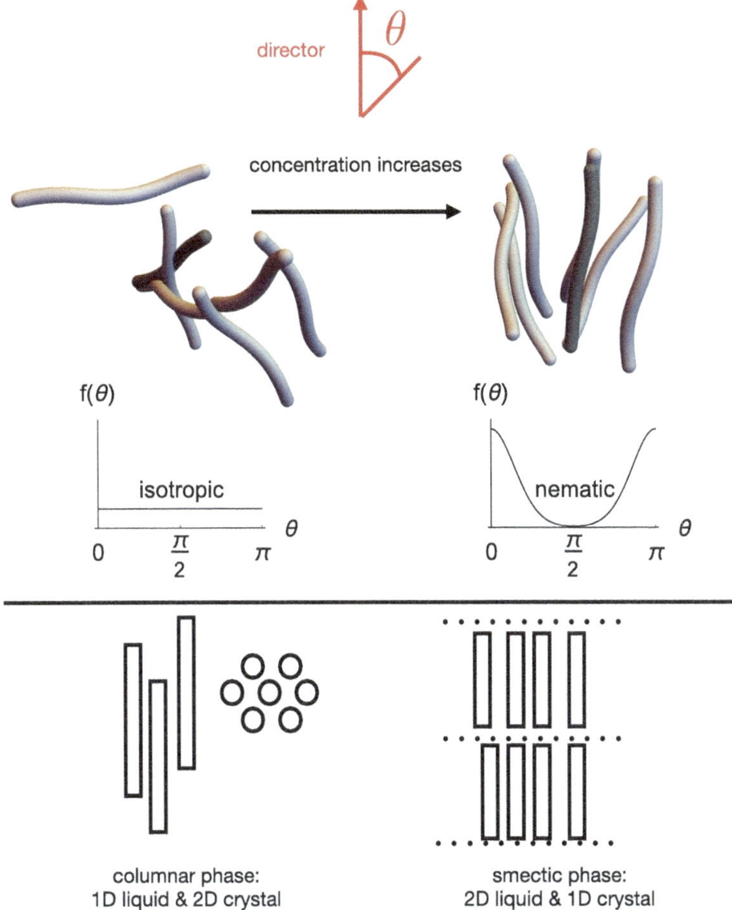

Fig. 5.25 Top: Randomly oriented rodlike polymers in solution at low concentration on the left versus the same polymers at a higher concentration on the right. Increasing crowding forces them to preferentially align. Middle: Sketches of the orientation distribution function $f(\theta)$ in the two cases, i.e. in the isotropic phase and the nematic phase, respectively. Bottom: Cartoons depicting additional liquid crystalline phases, i.e. a hexagonal columnar and a smectic phase

do possess temperate dependence as well—either via the temperature dependence of their bending rigidity or via 'soft' interactions. Since the persistence length P is important for the relative stability of the liquid crystalline phases formed by rodlike polymers, we briefly digress and discuss the temperature dependence of P.

We may express the partition function of a persistent flexible polymer via

$$Q = \int \left(\frac{d\Omega}{4\pi}\right)^N \exp[-\beta H], \tag{5.35}$$

where

$$H = -J \sum_i \vec{t}_i \cdot \vec{t}_{i+1} \quad (J > 0). \tag{5.36}$$

Let's momentarily assume that J is a constant. Hence

$$Q = \Pi_{i=1}^N \int_0^\pi \frac{d\theta_i \sin\theta_i}{2} \exp[J\cos\theta_i] = \left(\frac{\sinh\beta J}{\beta J}\right)^N. \tag{5.37}$$

Here the quantities θ_i are the angles between neighboring tangent vectors along the contour. From (5.37) we find $\langle\cos\theta\rangle$, i.e.

$$\langle\cos\theta\rangle = \frac{1}{N}\frac{\partial\ln Q}{\partial(-\beta J)} = \coth\beta J - \frac{1}{\beta J} \approx 1 - \frac{1}{\beta J} \quad (\beta J \gg 1). \tag{5.38}$$

Note that $\beta J \gg 1$ is the proper limit, because we want $\langle\cos\theta\rangle \approx 1 - \langle\theta^2\rangle/2$ (cf. (2.24)). Using (2.25) we find

$$P = \frac{bJ}{k_B T}. \tag{5.39}$$

Thus, $P \sim T^{-1}$. Note however, that this is a model in which bJ is a constant. In a real persistent flexible molecule bJ itself usually depends on temperature, which modifies $P \sim T^{-1}$ of course.

Equation (5.39) can be obtained quite differently using fluctuation theory. This approach is very insightful—as we shall see when we discuss the persistence length of polyelectrolytes. However, in order to avoid distraction from our current line of thought, the calculation of P from fluctuation theory is moved to Appendix B.

The following consists of three parts. First we want to study the statistical mechanics of the isotropic to nematic transition of monodisperse rods depending on number concentration ρ, aspect ratio $x = L/D$, where L and D are the length and the diameter of the rodlike polymers respectively, and flexibility L/P. Subsequently we want to discuss, without explicit calculations, how to modify our theory in order to accommodate other phases exhibiting orientational and translational order. Finally, we need to discuss applications.

We begin by deriving the orientation entropy,[1] expressed in terms of $f(\theta)$, of what we had called a persistent flexible or wormlike chain, i.e. a locally cylindrical object to which (2.22) applies. We use the **self-consistent field method**, where our starting point is (2.46):

$$
\begin{aligned}
Q_{n+1}(\vec{t}_1, \vec{t}_{n+1}) &= Q_n(\vec{t}_1, \vec{t}_{n+1}) \\
&+ \frac{1}{2} \int' \frac{d\Omega_n}{4\pi} \left(\vec{t}_n - \vec{t}_{n+1}\right)^2 \Delta_\Omega Q_n(\vec{t}_1, \vec{t})\big|_{\vec{t}_{n+1}} + \\
&- \beta u(\vec{t}_{n+1}) Q_n(\vec{t}_1, \vec{t}_{n+1}).
\end{aligned}
\tag{5.40}
$$

Here x_n is replaced by $\vec{t}_n$, which is the unit tangent vector of the worm's contour at the nth element of length b (cf. Fig. 2.8). The linear term in the expansion vanishes as before in (2.46). In addition, $\sum_{x_n}'$ is replaced with $\int' d\Omega_n/(4\pi)$. Note that the orientation of each tangent vector is defined by the angles φ and θ, i.e. $\vec{t} = \vec{t}(\varphi, \theta)$. θ_{n+1} is the angle between $\vec{t}_n$ and $\vec{t}_{n+1}$ and φ_{n+1} is the rotation angle of $\vec{t}_{n+1}$ around the direction defined by $\vec{t}_n$. Hence $d\Omega_n = \sin\theta_n d\theta_n d\varphi_n$. Finally, $\Delta_\Omega = (\sin\theta)^{-2}\partial_\varphi^2 + (\sin\theta)^{-1}\partial_\theta(\sin\theta\,\partial_\theta)$ is the angular part of the Laplace operator. We evaluate the integration as follows:

$$
\int' \frac{d\Omega_n}{4\pi} \left(\vec{t}_n - \vec{t}_{n+1}\right)^2 =
\tag{5.41}
$$

$$
2\int' \frac{d\Omega_n}{4\pi}(1 - \cos\theta_n) \approx \int' \frac{d\Omega_n}{4\pi}\theta_n^2 = \frac{1}{2}\langle\theta_n^2\rangle.
$$

Here we have used that the local bending of the contour is small. And note that, as the prime on the integral indicates, there is only one general direction for $\vec{t}_n$, which is responsible for the factor $1/2$. Similarly we may expand the two sides in (2.22), which yields

$$
\langle\theta_n^2\rangle \approx \frac{2b}{P}.
\tag{5.42}
$$

Combination of the last three equations and following the step from (2.47) to (2.48) yields

$$
\frac{dQ_L(\vec{t}', \vec{t})}{dL} = \frac{1}{2P}\Delta_\Omega Q_L(\vec{t}', \vec{t}) - \beta\frac{u(\vec{t})}{b}Q_L(\vec{t}', \vec{t}),
\tag{5.43}
$$

---

[1] At this point 'conformation entropy' may be the better expression. What this has to do with orientation will become clear as we move along.

where $L = nb$.

Since (5.43) is the equivalent of (2.48), we find

$$-\frac{\mu_o}{b}\psi_o(\vec{t}) = \frac{1}{2P}\Delta_\Omega\psi_o(\vec{t}) - \beta\frac{u(\vec{t})}{b}\psi_o(\vec{t}) \tag{5.44}$$

and

$$\frac{\Delta F_{orient}}{Lk_B T} \approx \frac{1}{b}\mu_o. \tag{5.45}$$

Hence the conformational entropy becomes

$$\frac{\Delta S_{orient}}{k_B} = \frac{\Delta E_{orient}}{k_B T} - \frac{\Delta F_{orient}}{k_B T} = \frac{L}{b}\int\frac{d\Omega}{4\pi}f(\vec{t})\beta u(\vec{t}) - \frac{L}{b}\mu_o. \tag{5.46}$$

Note that $f(\vec{t})$ is the orientation probability distribution of $\vec{t}$. It plays the role of $c(x)$ in (2.53) and thus $f(\vec{t}) = \psi_o^2(\vec{t})$. If we use this in conjunction with (5.44), the conformational entropy becomes

$$\frac{\Delta S_{orient}}{k_B} = \frac{L}{2P}\int\frac{d\Omega}{4\pi}\psi_o(\vec{t})\Delta_\Omega\psi_o(\vec{t}). \tag{5.47}$$

Our worms possess a cylindrical cross section and thus $f(\vec{t}) = f(\theta)$. Hence

$$\frac{\Delta S_{orient}}{k_B} = \frac{L}{2P}\int\frac{d\theta}{2}\psi_o(\theta)\partial_\theta\sin\theta\partial_\theta\psi_o(\theta) \tag{5.48}$$

$$\stackrel{p.i.}{=} -\frac{L}{2P}\int\frac{d\Omega}{4\pi}(\partial_\theta\psi_o(\theta))^2,$$

where p.i. stand for 'partial integration'. We obtain ΔS_{orient} in its final form by replacing $\psi(\theta)$ with $\sqrt{f(\theta)}$, i.e.

$$\frac{\Delta S_{orient}}{k_B} = -\frac{L}{8P}\int\frac{d\Omega}{4\pi}\left[\partial_\theta f(\theta)\right]\left[\partial_\theta\ln f(\theta)\right]. \tag{5.49}$$

There is another limit, which is of interest here. In this limit $P \gg L$, which means that the worms essentially are rigid cylinders. What is ΔS_{orient} in this limit? The classical partition function of an N-point particle gas contains a factor $N!^{-1}$, which accounts for the indistinguishability of the particles. If the particles come in groups i $(= 1, 2, \dots)$ and if only the N_i particles belonging to the same group are indistinguishable, then this factor becomes $\Pi_i N_i!^{-1}$ instead, where $\sum_i N_i = N$.

Here the particles in group i are the cylinders whose axes lie within the solid angle $d\Omega_i$ and $N_i/N = f(\theta_i)$. With this we find

$$\ln \Pi_i N_i!^{-1} = N \sum_i \left(-\frac{N_i}{N} \ln \frac{N_i}{N} + \frac{N_i}{N} \right) - N \ln N \tag{5.50}$$

$$= -N \int \frac{d\Omega}{4\pi} f(\theta) \ln f(\theta) - N \ln N + N.$$

Note that we have made use of Stirling's formula (3.3). Obviously, the integral accounts for the entropy tied to the orientation distribution of the particles, i.e. the cylinders. Specifically,

$$\frac{\Delta S_{orient}}{k_B} = -\int \frac{d\Omega}{4\pi} f(\theta) \ln f(\theta) \tag{5.51}$$

is the equivalent to (5.49) in this limit.

A formula for the orientation entropy joining the two limits (5.49) and (5.51) was derived by Khokhlov and Semenov [46] based on work by I. M. Lifshitz (see also [47]). For the purpose of this section, however, the limiting expressions (5.49) and (5.51) are sufficient.

In order to find out whether there is a phase transition from an isotropic to a nematic phase, we need the entire free energy of the system. In particular we need to include the interaction between the polymers. For this purpose we return to the first and prototypical calculation of this type carried out by Onsager [48]. Onsager's free energy for hard rigid cylinders has the form

$$\frac{F}{Nk_B T} = \frac{\mu_o}{k_B T} + \sigma(f) + \ln \rho - 1 + x \phi g(f). \tag{5.52}$$

The first term lumps together all contributions to the free energy which do not depend on the orientation and/or concentration of the rodlike particles. $\sigma(f)$ is simply $-\Delta S_{orient}/k_B$ in (5.51). In the next term, $\rho = N/V$ is the number density of the cylinders. In the last term $x = L/D$ is the length of the cylinder divided by its diameter and $\phi = x\,b$, where b is the volume of a cylinder. This term, which corresponds to the second order in a virial expansion, describes the excluded volume between the cylinders. The function $g(f)$ is given by

$$g(f) = 64\pi \int \int \frac{d\Omega \, d\Omega'}{4\pi \; 4\pi} f(\theta) f(\theta') |\sin(\Omega - \Omega')|, \tag{5.53}$$

where $\Omega - \Omega'$ denotes the angle between the axes of two interacting cylinders. Note that $DL^2|\sin(\Omega - \Omega')|$ is the volume which one cylinder excludes to the center of mass of the other (remember the geometrical meaning of the cross product between to vectors). Onsager did his calculation for spherocylinders, i.e. cylinders with hemispherical endcaps, in order to avoid the complexity of the 'naked' cylinder ends.

It is useful to describe $f(\theta)$ via a trial function, which is close to the expected shape of the true orientation distribution function. Onsager chooses

$$f(\theta) = \frac{\alpha}{\sinh(\alpha)} \cosh(\alpha \cos \theta). \tag{5.54}$$

The quantity α is a parameter. It is equal to zero in the isotropic phase and thus the orientation distribution function is constant. In the nematic phase it exhibits maxima at $\theta = 0$ and $\theta = \pi$, respectively. Using this trial function we obtain for $\sigma(f)$

$$\sigma(f) = \begin{cases} 0 & \alpha = 0 \\ \ln \alpha - 1 & \alpha \to \infty \ (P \gg L) \\ \frac{L}{4P}(\alpha - 1) & \alpha \to \infty \ (P \ll L) \end{cases}. \tag{5.55}$$

Note that we still concentrate on the limit $P \gg L$, even though here we have added $\sigma(f)$ in the opposite limit (cf. (5.49)) for later use! In addition, Onsager obtains an analytic solution for the interaction integral. This is a somewhat tedious calculation, but it is straightforward to obtain the limiting expressions:

$$g(f) = \frac{2I_2(2\alpha)}{\sinh^2 \alpha} = \begin{cases} 1 & \alpha = 0 \\ 4(\pi\alpha)^{-1/2} & \alpha \to \infty \end{cases}. \tag{5.56}$$

The quantity I_2 is a modified Bessel function (e.g., [49]; Sect. 9.6).

In an anisotropic phase (index a) the parameter α must satisfy $dF_a/d\alpha = 0$, which yields

$$\alpha_a = \frac{4}{\pi} x^2 \phi_a^2. \tag{5.57}$$

The coexisting volume fractions ϕ_i and ϕ_a along an isotropic to anisotropic (nematic) phase boundary follow from the conditions that the **osmotic pressure**,

$$\Pi = -\frac{dF}{dV} = k_B T b^{-1} \phi \left(1 + x\phi g(f)\right), \tag{5.58}$$

and the chemical potential,

$$\mu = \frac{dF}{dN} = \mu_o + k_B T \left(\sigma(f) + \ln(\phi/b) + 2x\phi g(f) \right), \tag{5.59}$$

are the same on the two sides of the transition (Question: In (5.58) the osmotic pressure is calculated via the negative derivative of the free energy with respect to volume. Why can this be the osmotic pressure? To answer this question, compare the (3.19) and (3.38) (setting $m_2 = 1$).). Setting $\Pi_i = \Pi_a$ and $\mu_i = \mu_a$ leads to

$$\phi_i'(1 + \phi_i') = 3\phi_a' \tag{5.60}$$
$$\ln \phi_i' + 2\phi_i' = 3\ln \phi_a' + 3 + \ln(4/\pi), \tag{5.61}$$

where $\phi_i' = x\phi_i$ and $\phi_a' = x\phi_a$. These equations possess the stable solution

$$\phi_i \approx 3.45 \frac{D}{L} \quad , \quad \phi_a \approx 5.12 \frac{D}{L} \quad , \text{ and } \quad \alpha_a \approx 33 \tag{5.62}$$

(the exact numerical values of the numbers multiplying D/L are 3.29 and 4.19, respectively). Note that the large value of α_a justifies the assumption of large α in the anisotropic phase.

Ordinarily, the second virial approximation of the particle-particle interaction is not sufficient for this type of calculation, i.e. the calculation of densities at coexistence. Here, however, we assume that x is so large that both ϕ_i and ϕ_a are sufficiently small to permit the use of the second virial approximation with reasonable accuracy. The experimental system which Onsager originally had in mind is tobacco mosaic virus (TMV). TMV is a highly elongated, stiff virus particle 300 nm in length and 18 nm wide ($x \approx 17$) composed of over 2000 protein segments. Figure 5.26 shows an electron micrograph image of TMV.

It is worth noting that Onsager's inspiration most likely did not come from possible industrial applications of **liquid crystalline polymers**—this type of research gained traction in the 1960s with the advent of liquid crystal display technology and high strength polymer fibers. He was interested in the physics of phase transitions in general.

Let us now study the other limit, i.e. $L \gg P$. The free energy we use in his case is again (5.53) with $\sigma(f)$ now being the second expression in (5.55), i.e.

$$\frac{F}{Nk_B T} = \frac{\mu_o}{k_B T} + \frac{L}{P} \frac{\alpha - 1}{4} + \ln(\rho_p/b_p) - 1 + \frac{L}{P} x \phi_p g(f), \tag{5.63}$$

where $b_p = \frac{\pi}{4} D^2 P$ and $\phi_p = b_p \rho$. We want to study the limit $L/P \to \infty$ and this form of the free energy highlights the dominant terms in this limit. In addition we assume that the excluded volume interaction remains what it was before. This is not really correct, but the approximation has proven useful. As before we minimize

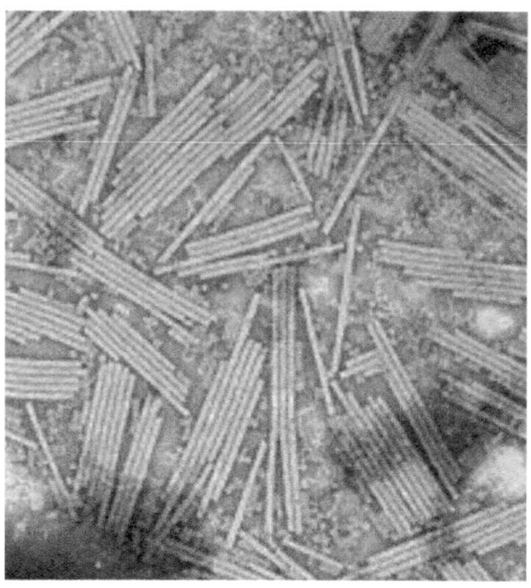

Fig. 5.26 Electron micrograph of TMV

the free energy with respect to α and work out the osmotic pressure as well as the chemical potential. The result is

$$-\frac{1}{4} = \tilde{\phi}_a \, g'(\alpha_a) \tag{5.64}$$

$$\tilde{\phi}_i^2 = \tilde{\phi}_a^2 \, g(\alpha_a) \tag{5.65}$$

$$\tilde{\phi}_i = \tilde{\phi}_a \, g(\alpha_a) + \frac{\alpha_a - 1}{8}, \tag{5.66}$$

with $\tilde{\phi} = \frac{P}{D}\phi$, $g(\alpha) = 4/\sqrt{\pi\alpha}$, and $g'(\alpha) = -2/\sqrt{\pi\alpha^3}$. Now the numerical solution is

$$\phi_i \approx 7.12\frac{D}{P} \quad , \quad \phi_a \approx 8.75\frac{D}{P} \quad , \text{ and } \quad \alpha_a \approx 11.6. \tag{5.67}$$

The value we find for α_a is not very large, which means that our solution is not accurate. We also need $D \ll P$ in order for the transition to occur at sufficiently small volume fractions. However, here we are interested in the qualitative picture. Most importantly we learn that the coexistence is shifted to higher volume fractions with increasing flexibility, i.e. smaller persistence length.

Figure 5.27 is an application of what we have discussed thus far. The osmotic pressure data were obtained for PBLG in an organic solvent—dimethylformamide (DMF). The dashed line is Onsager's theory as we have discussed it except for one difference. Here the authors have not used the approximations for large α. Instead they compute the free energy, the osmotic pressure as well as the chemical potential

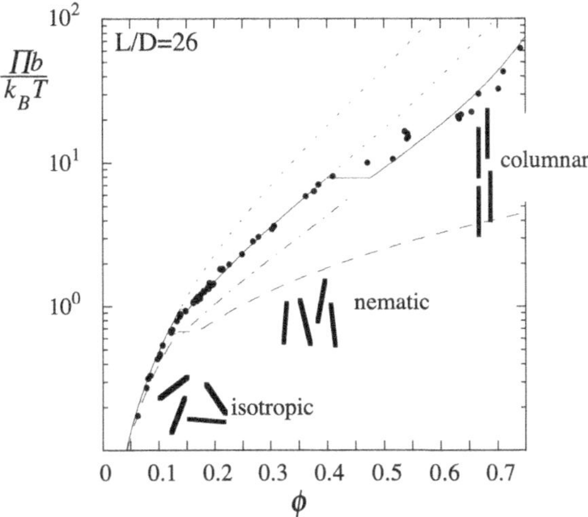

Fig. 5.27 Experimental and theoretical reduced osmotic pressure of PBLG in DMF versus solute volume fraction. This is Fig. 3 from [50]. Here b is the molecular volume. Solid line: Theory based on the Khokhlov-Semenov approach to flexibility in combination with the decoupling approximation and the dimensional separation approach based on Eq. (8) in [51]. Upper dotted line: Extension of the isotropic branch. Lower dotted line: Extension of the nematic branch. Dashed-dotted line: Result for completely rigid molecules. Dashed line: Onsager's 2nd virial coefficient-approximation for rigid rods

numerically, but nevertheless based on the above trail function. The solid line, which is a much better fit to the data, includes the flexibility of PBLG. Again the authors do not use the limiting forms of $\sigma(f)$. Instead they use the aforementioned interpolation formula, which approximates $\sigma(f)$ for arbitrary L/P. Here the persistence length ($P \approx 200\,\text{nm}$) is an adjustable parameter.

In addition to the isotropic-to-nematic transition, the figure also includes a third phase, i.e. a columnar phase. Avoiding all details, which can be found in [51], the underlying idea is quite straightforward. In a columnar phase, the part of the free energy modelling the excluded volume interaction is assumed to be the sum of two terms, i.e.

$$F_{ex.vol.}(\Delta) = F^{1D}_{fluid}(\Delta) + F^{2D}_{crystal}(\Delta). \qquad (5.68)$$

$F^{1D}_{fluid}(\Delta)$ is the excluded volume part of the free energy of a one-dimensional hard body fluid. $F^{2D}_{crystal}(\Delta)$ is the analogous part of the free energy of a two-dimensional system of hard disks exhibiting hexagonal ordering. The quantity $F^{1D}_{fluid}(\Delta)$ is realized via the so-called **scaled particle theory** (see for instance [52]), whereas $F^{2D}_{crystal}(\Delta)$ is realized using a **cell model** (see for instance [53]). What combines these two expressions into the three-dimensional $F_{ex.vol.}(\Delta)$ is the quantity Δ. This

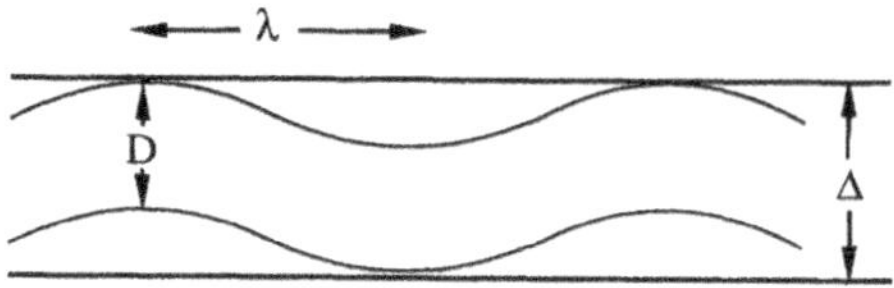

Fig. 5.28 Cartoon of a persistent flexible polymer inside a tube of diameter Δ. The quantity D is the diameter of the polymer and λ is the so-called deflection length explained in the text. Reprinted with permission from [51]

quantity is depicted in Fig. 5.28. It is the width of a tube to which the polymer is confined. Viewed along the direction of the polymer backbone, it is the width of the aforementioned 2D cell (a hexagon in this particular case). Note that the 3D polymer volume fraction ϕ therefore is a unique function of Δ as well. Note also that Δ in conjunction ϕ determines the 1D polymer volume fraction in $F_{fluid}^{1D}(\Delta)$, i.e. this is why F_{fluid}^{1D} also depends on Δ. For any fixed value of ϕ the most stable free energy is found via minimization of the 3D free energy with respect to Δ.

If the polymer is flexible, then there also exists a relation between Δ and α. According to Fig. 5.28 the polymer contour winds its way along the tube and is defected by the tube's wall every so often. The mean distance between successive deflections is λ, which is called the **deflection length** (a term originally coined by T. Odijk). Based on this picture we construct a relation between Δ and λ via

$$\Delta - D \sim \int_0^\lambda \langle \theta(s)^2 \rangle^{1/2} ds \sim \frac{\lambda^{3/2}}{P^{1/2}}, \tag{5.69}$$

where we use $\langle \theta(s)^2 \rangle \sim s/P$ according to $\langle \vec{t}(0) \cdot \vec{t}(s) \rangle = \langle \cos \theta(s) \rangle \approx 1 + \frac{1}{2}\langle \theta(s)^2 \rangle \approx 1 + s/P$. Similarly we make use of $\langle \theta^2(\lambda) \rangle \sim \lambda/P$ and $\langle \theta^2(\lambda) \rangle \sim \alpha^{-1}$, i.e.

$$\lambda \sim \frac{P}{\alpha}. \tag{5.70}$$

Putting everything together we have

$$\alpha = \kappa [P/(\Delta - D)]^{2/3}, \tag{5.71}$$

where $\kappa = O(1)$ is an undetermined (and thus adjustable) proportionality constant.

Figure 5.29 shows the phase behavior of persistent flexible polymers with hard-body excluded volume interactions as a result of this theory. Independent of the value of κ, the general phase behavior is the following. Stiff polymers undergo a transition from isotropic (I) to nematic (N) at a volume fraction determined by their aspect ratio, i.e. the larger L/D becomes the smaller ϕ_i and ϕ_a will be. At significantly higher

Fig. 5.29 Phase diagrams for persistent flexible polymers for three different axial ratios and continuously variable persistence length. Solid lines: $\kappa = 3.5$; dashed lines $\kappa = 2.5$. Reprinted with permission from [51]

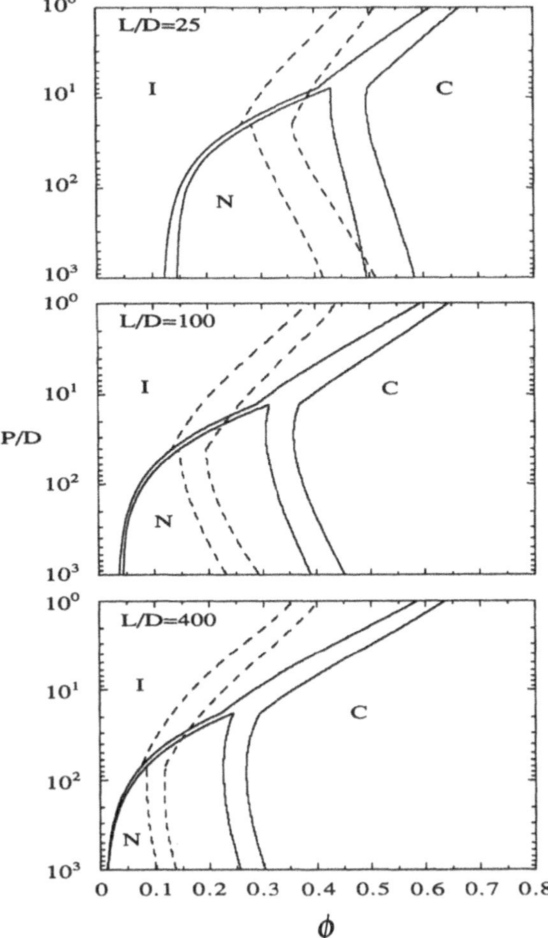

volume fractions a transition into a columnar (C) phase occurs. With increasing flexibility the transition from I to N is postponed to larger ϕ until finally the nematic phase is squeezed out between the I and C phases.

It is important to note that this theoretical approach to translational ordering, intuitive though it is, has one significant drawback. Since the free energies of these orderings are constructed as separate branches, the resulting transition will always be a first order transition even if the true transition is not first order.

Thus far we have dealt with hard-body or athermal interactions between rod-like flexible polymers. A rich and different phase behavior is encountered for soft interactions. Figure 5.30 shows an example taken from the book by Grosberg and Khokhlov. Here the interaction between the rodlike molecules, possessing a small aspect ratio, is attractive. The resulting phase diagram resembles the gas-liquid-solid

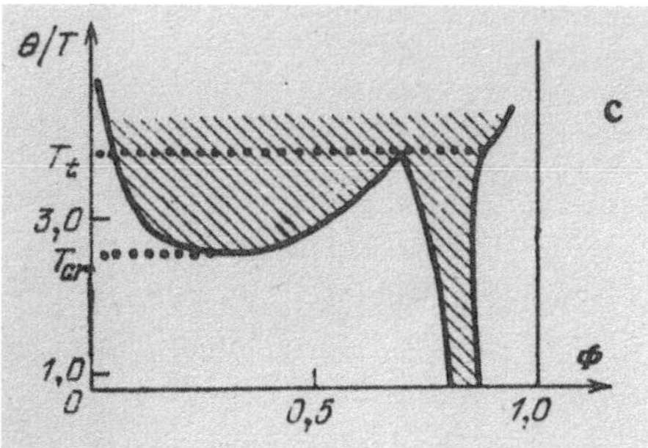

Fig. 5.30 Theoretical phase diagram, inverse reduced temperature versus polymer volume fraction, of a liquid crystalline solution of rodlike polymers with a short aspect ratio. θ is the θ-temperature. The shaded region indicates phase coexistence. Excerpt from Fig. 5.4 in the book by Grosberg and Khokhlov

phase diagram of a simple liquid, except that the narrow phase coexistence separates an isotropic fluid from a nematic liquid at fairly high volume fraction. When the aspect ratio is large, a similar phase diagram is obtained. In this case the narrow phase coexistence funnel appears at low volume fractions. In addition, the one phase region bracketed between the critical temperature T_{cr} and the triple point temperature T_t as well as between the funnel and the broad coexistence region is not an isotropic fluid as before but a nematic—albeit less ordered than the nematic phase on the far side of the funnel. What is not shown here, but is discussed in the aforementioned book, is what happens to this phase diagram in the crossover region between small and large aspect ratios or what happens as function of the persistence length.

All in all the above is only a very small fraction of what has been studied. Important aspects include not only different types of interactions but also polydispersity, molecular shape, or, more generally, molecular architecture (e.g., instead of the macromolecular contour, on which we have concentrated here, much smaller side groups tethered to the backbone may be the actual mesogens causing liquid crystalline ordering). Since this is not the place for an in depth discussion, a few references must suffice. Early but nonetheless valuable references include [54–56]. A comprehensive discussion of the subject can be found in [57]. Quite recently there is renewed interest from the computer simulation community in this subject. See for example [58] and references therein.

We have not yet mentioned applications of liquid crystalline polymers. Here are a few: ropes and cables, electronic support structures, recreation and leisure industry, aerospace and military, new textiles, industrial applications. One material property of particular importance is **tensile strength** (which can easily surpass that of steel). Tensile strength is greatly enhanced if one succeeds in aligning the polymer backbones in the direction of the force (Kevlar perhaps is the best-known example). This is the principle underlying the high strength polymer fibers can exhibit. When such fibers are spun from the melt or a solution, the flow field in the nozzle already tends to promote alignment of the stiff backbones. Subsequently the fiber is stretched. The amount of stretching is described by the so-called **spin stretch factor**. Again, the crowding of the backbones enhances alignment. As a result, the contour length of the molecules approximates their end-to-end distance. Hence, the elongation at break of such a fiber is only a few percent. While these industrial manufacturing processes are complicated in detail, theoretical concepts like the ones just outlined help to understand and optimize them. It is worth noting that the pronounced anisotropy of high strength fibers has an important negative consequence. While the tensile strength is greatly enhanced by molecular alignment, the **compressive strength** is significantly reduced, i.e. the fiber's shear modulus is small. The tensile strength of a high strength fiber tied into a knot therefore is very much reduced. An informative reference in this context is [59].

Reversibly Assembling Liquid Crystalline Polymers:

Thus far, the **mesogen**, i.e. a compound that displays liquid crystal properties, does not change, e.g., as a function of concentration. This is quite different in so-called **self-assembling systems**, where the molecules form labile **supramolecular structures** above a critical concentration—the **critical micelle concentration** or CMC. Labile means that these structures are not held together by covalent bonds, but by much weaker van der Waals, hydrophobic, hydrogen bridge or electrostatic interactions. Hence they can reversibly assemble and disassemble in response to certain control parameters, e.g. concentration. The resulting supramolecular structures include, for example, spheres or cylinders (compact forms are usually referred to as **micelles** or **vesicles**), **lamellar bilayers** or complex sponge-like structures (see for instance [29, 60]). Which structures are observed, or whether the existing supramolecular structure transforms into a different new supramolecular structure, depends on numerous variables in addition to the concentration, such as the temperature, the composition (in multicomponent systems), the ionic strength, etc. In the following we shall focus on **reversibly aggregating amphiphilic systems**, forming simple compact shapes that can be described as flexible rods. Note that **amphiphiles** are bi-polar molecules, such as surfactants or lipids, composed of a **hydrophobic** and a **hydrophilic** part. This means that in aqueous solution there is a tendency for the hydrophobic parts to cluster together, as much as sterically possible, in order to reduce the unfavorable interaction with water (**hydrophobic effect**).

In the case of shape-anisotropic aggregates one can easily imagine that the excluded volume interaction between them can lead to the formation of liquid crystalline mesophases. And indeed, many self-aggregating amphiphilic systems exhibit mesophases characterized by different types of ordering of their aggregates, e.g., nematic, smectic, columnar, or crystalline (see for instance [61]). However, since the underlying mesogen is no longer a stable unit, there is an interesting coupling tying the variability of the mesogen to the phase behavior. Before we progress further along this direction, it is useful to summarize the basic principles of micellar assembly.

Figure 5.31 is a sketch of amphiphilic molecules reversibly assembling into spherical micelles. Here the head groups, shown as filled black circles, preferentially mix with water, whereas the tails, depicted as zigzag-lines, do not like to mix with water. This leads to a clustering of the zigzag-tails into droplets shielded on the outside by the head groups. Clearly, here we are interested in the assembly leading to aggregates with pronounced shape anisotropy. This is because we want to study the type of liquid crystalline ordering which we have discussed already in the preceding section.

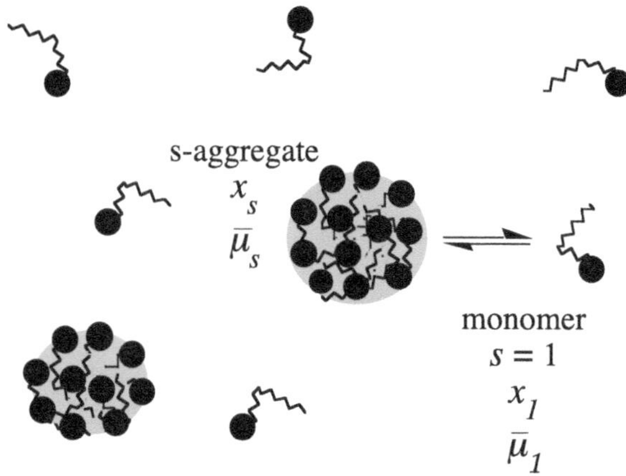

Fig. 5.31 Illustration of the reversible assembly of amphiphilic molecules. Here the aggregates are spherical micelles. Reprinted with permission from [52]

Nevertheless, the following thermodynamic considerations are independent of the aggregate shape.

The starting point is the reaction equilibrium

$$s A_1 \rightleftharpoons A_s. \tag{5.72}$$

A_s denotes a s-aggregate containing s monomers A_1. Expressed in terms of the chemical potentials this becomes

$$s \mu_1 = \mu_s. \tag{5.73}$$

Assuming that the solute concentration is low, this can be expressed as

$$s \bar{\mu}_1 + s R T \ln x_1 = \bar{\mu}_s + R T \ln(x_s/s). \tag{5.74}$$

Note that the quantity x_s/s is the mole fraction s-aggregates and therefore x_s is the mole fraction of monomers bound in s-aggregates. Solving (5.74) for x_s yields

$$x_s = s \, (x_1 e^\alpha)^s, \tag{5.75}$$

where

$$\alpha = \frac{1}{RT} \left(\bar{\mu}_1 - \frac{1}{s} \bar{\mu}_s \right). \tag{5.76}$$

Equation (5.75) has an interesting consequence. Note that the total monomer mole fraction is given by

$$x = x_1 + \sum_{s=m}^{\infty} x_s.$$ (5.77)

Here x_1 is the mole fraction of the free monomers, whereas the sum is equal to the mole fraction contributed by the monomers bound in aggregates. The quantity m is a minimum aggregate size. In the case of spherical micelles it results form the requirement that a certain number of head groups are needed to form a closed surface. This number may be large—say $m \approx 50$—depending of course on the type of monomer. But $m = 2$ is also possible, for instance in the case of stacking amphiphiles forming rodlike aggregates (see for example Boden in [62]). Since $x \leq 1$, the right hand side of (5.77) is bounded. This implies that $x_1 e^{\alpha} < 1$, because the sum $\sum_{s=m}^{\infty} sq^s$ diverges at $q = 1$ (geometric series!). Hence,

$$x_{CMC} = e^{-\alpha}$$ (5.78)

defines a threshold mole fraction, which the free monomers cannot exceed. Instead, they condense into aggregates if the overall solute mole fraction increases beyond x_{CMC}. The corresponding concentration is the critical micelle concentration or, more generally, the **critical aggregate concentration**. The sharpness of this condensation 'transition' is governed by the size of m, i.e. increasing m sharpens the onset of aggregate formation.

Let's conclude this discussion by studying the dependence of the average size $\langle s \rangle$ of linear aggregates, i.e. $m = 2$, on the overall solute mole fraction x. This requires a model for $\bar{\mu}_s$. A simple one is

$$\bar{\mu}_s = s\bar{\mu}_1 - RT\alpha_o(s - 1).$$ (5.79)

The quantity $-RT\alpha_o$ is a contact free enthalpy for each of the $s - 1$ contacts between neighboring monomers in s-aggregates. With this we have $\alpha = \alpha_o(s - 1)/s$. It is straightforward to work out

$$\langle s \rangle = \frac{\sum_{s=1}^{\infty} sx_s}{\sum_{s=1}^{\infty} x_s} = \frac{1+q}{1-q} \quad \text{and} \quad x = \sum_{s=1}^{\infty} x_s = e^{-\alpha_o} \frac{q}{(1-q)^2},$$ (5.80)

where $q = x_1 e^{\alpha_o}$. Solving the left equation for q and inserting the result into the right equation yields

$$\langle s \rangle = \sqrt{1 + 4e^{\alpha_0}x}.\tag{5.81}$$

We find that the mean aggregate size increases proportional to $\sqrt{x}$ for sufficiently large $\langle s \rangle$. This aggregate growth in the ideal system is enhanced when the aggregates interact (cf. Gelbart and Ben-Shaul [62]).

Theoretical studies of the statistical mechanics of reversibly aggregating systems can be divided roughly into three categories. The first mainly comprises the formation of the isolated micelles and their internal structure at low concentrations (just above the CMC), neglecting the interaction of the aggregates. The second comprises the phase theory of more (than two) component systems (microemulsions) by means of mostly lattice or field theoretical models (e.g., [63]; p. 26.). The third group, which is the subject here, involves the study of the liquid crystalline behavior of the labile polydisperse aggregate population at higher solute concentrations by means of a synthesis of the excluded volume models discussed above with phenomenological descriptions of the amphiphilic interactions (cf. [61]).

In the following we consider the above case of monomers forming linear (and rigid) aggregates. A simple expression for the free energy change during the transition from the isotropic to the nematic phase in such a system is given by

$$\frac{F}{Vk_BT} = \sum_{s=1}^{\infty} \int \frac{d\Omega}{4\pi} f(s,\theta) \ln f(s,\theta)$$
$$+ \frac{1}{2} \sum_{s,s'=1}^{\infty} \int \int \frac{d\Omega\, d\Omega'}{4\pi\, 4\pi} f(s,\theta) f(s',\theta') b^{ex}(s,s',\Omega-\Omega')$$
$$+ \tilde{f}^{trans/rot} + \tilde{f}^{aggregate}.\tag{5.82}$$

This is a generalization of (5.52). The generalization is that the population of rods or aggregates is variable polydisperse, i.e. $\rho f(\theta) \to \sum_{s=1}^{\infty} f(s,\theta)$. Here $f(s,\theta)$ is the number of aggregates consisting of s monomers with orientation θ in the volume V. ρ is the aggregate number density. The $g(f)$-term in (5.52) is replaced by the b^{ex}-term. As in the Onsager model, the aggregates are absolutely stiff, hard rods, whose interaction again is modelled via their 2nd virial coefficient. $\tilde{f}^{trans/rot}$ stands for the translational free energy term, which in (5.52) merely is $\ln \rho - 1$, including the equivalent of $\mu_o/(k_BT)$ accounting for the vibrational and rotational degrees of freedom. These contributions appear because, unlike in the case of inert rods, the length distribution of the aggregates in the isotropic and in the nematic phase is different. Originally, these terms were considered important to prevent 'catastrophic' growth of the aggregates in the nematic phase. It is now known that this unphysical growth is prevented if flexibility [64–66] or a short-range soft repulsion [67] is taken into account in addition to the hard-body repulsion. The last term is an internal free energy of the aggregates, the simplest description of which is a negative temperature-dependent free energy per monomer-monomer contact $-k_BT\,\Phi(T)$, i.e.

$$\tilde{f}^{aggregate} = -\Phi(T) \sum_{s=1}^{\infty} \int \frac{d\Omega}{4\pi} f(s, \theta)(s - 1) \tag{5.83}$$

(cf. (5.79)). If one wishes to be more precise, or if one is dealing with more complicated aggregates, this expression must of course be modified accordingly, and $\Phi(T)$ is not only a function of T but also a function of, for example, the monomer position within the aggregate and/or the local curvature of the aggregate (or micelle) surface, as well as the surrounding solvent, etc. However, let's refrain from a more detailed discussion, because the exact treatment of these terms in this context is quite subtle.

Using the equilibrium condition $\mu_s = s\mu_1$, where $\mu_s = \partial(F/V)/\partial f(s, \theta)|_{V,T,f(s',\theta')}$ is the chemical potential of an aggregate of s monomers with orientation θ, it follows that

$$f(s, \theta) = \left(f(1, \theta)e^{\Phi - \chi(s)/s}\right)^s e^{-(\Phi - \chi(1))}. \tag{5.84}$$

The function $\chi(s) = 2\pi \sum_{s'=1}^{\infty} \int \frac{d\Omega'}{4\pi} f(s', \theta')b^{ex}(s, s', \Omega - \Omega') + \partial(\tilde{f}^{trans/rot} + \tilde{f}^{aggregate})/\partial f(s, \theta)|_{V,T,f(s',\theta')}$ is the contribution of the external interaction as well as the various degrees of freedom mentioned above. For general s, (5.84) must be solved numerically. However, if we first assume that $\chi(s)$ is negligible (for the external interaction this is the case at low amphiphile concentrations), then, with the help of the additional condition $\phi = b_1 \sum_{s=1}^{\infty} \int \frac{d\Omega}{4\pi} sf(s, \theta)$, where b_1 is the monomer volume, we can easily calculate the mean aggregation number $\langle s \rangle$ as a function of the amphiphile volume fraction ϕ, i.e.

$$\langle s \rangle \sim \left(\phi e^{\Phi}\right)^{1/2} \tag{5.85}$$

(cf. (5.81)).

Analogous to the Onsager model, this model can also be extended to isotropic-nematic phase behavior of linear aggregates at higher concentrations, including flexibility of the aggregates as well as soft interactions. Likewise, possible translationally-ordered phases can be included as well. Since these calculations are more complicated than for inert particles, we shall not go into details. A comprehensive discussion of the different approaches and an overview of the literature can be found in [61]. Here a single example must suffice.

Figure 5.32 illustrates the molecular effects of sickle cell anaemia. Pauling, Itan and Ingram have shown that in this disease a genetic defect leads to a faulty amino acid sequence of the hemoglobin molecule and thus to the formation of a hydrophobic 'pocket'. Under certain conditions (low oxygen partial pressure), this leads to reversible aggregation of the hemoglobin molecules in the erythrocytes of sickle cell patients. The aggregates formed are stiff rods of approximately 180–240 Å in diameter, wherein the globular monomer has a size of approximately 60 Å. The aggregation and the associated nematic orientation of the aggregates (see Fig. 5.32

Fig. 5.32 Top: Image showing 'donut'-shaped normal red blood cells (size approximately $10\,\mu\mathrm{m}$). Middle: Red blood cells of a sickle cell patient distorted by bundles of aggregated hemoglobin. Bottom: Close-up image of a burst red blood cell with protruding rodlike aggregates composed of globular HbS protein. Pictures reproduced from [68]

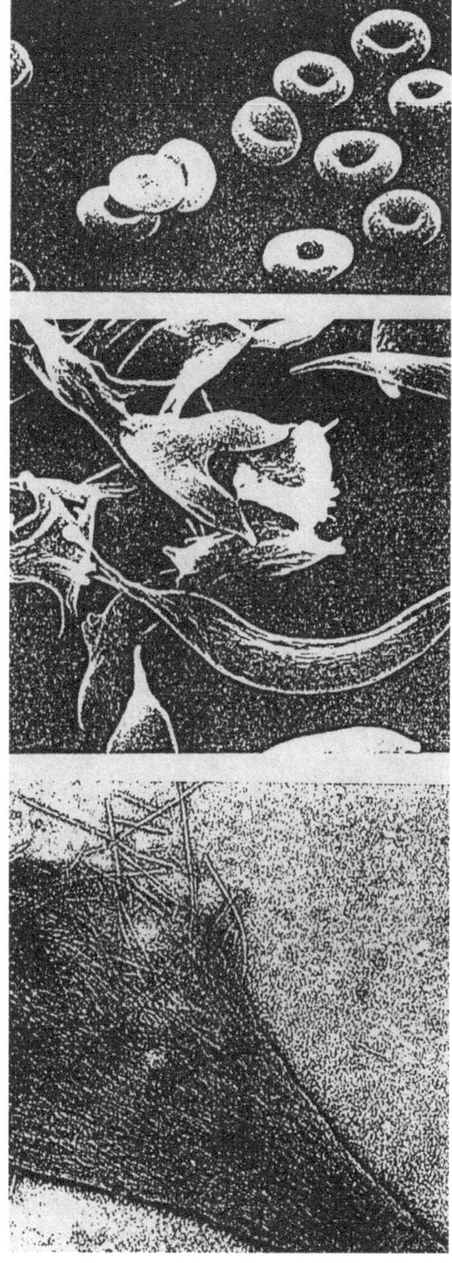

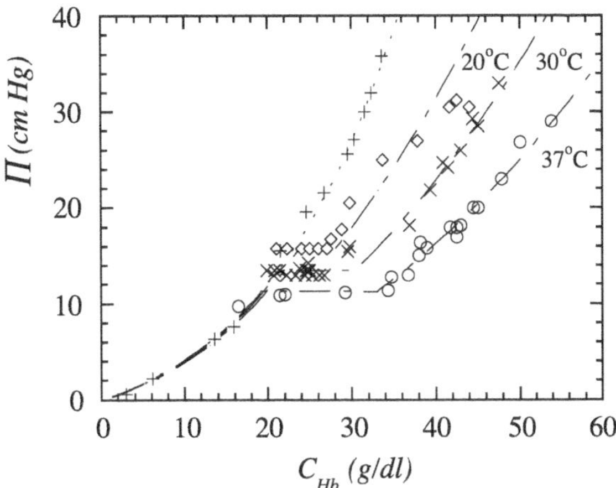

Fig. 5.33 Osmotic pressure versus concentration for deoxygenated hemoglobin. Reprinted with permission from [69]. Symbols: experimental data for normal hemoglobin (plusses) and sickle hemoglobin (37 °C (circles), 30 °C (crosses), and 20 °C (diamonds)). Dotted line: theoretical result for normal hemoglobin (T = 37 °C). Dashed-dotted lines: theoretical fits for highly concentrated deoxygenated sickle hemoglobin

(bottom)) leads to sickle-shaped deformation of the erythrocytes (see Fig. 5.32 (middle)), causing them to become rigid and thus hindering or preventing their capillary flow, i.e. oxygen transport. In addition, the erythrocyte membrane often is damaged (see Fig. 5.32 (bottom)). The theoretical description of this phenomenon via the osmotic equation of state, in particular across the isotropic-nematic transition, within the framework of an excluded volume theory is depicted in Fig. 5.33.

5.4 Polyelectrolytes

● *Preliminaries*

Polyelectrolytes are water-soluble polymers that carry dissociable groups along their contour. An example is poly(sodium styrene sulfonate) (PSS) shown in Fig. 5.34. PSS is a synthetic polyelectrolyte. However, many biological molecules are also polyelectrolytes. A famous example is DNA shown in Fig. 5.35. Polyelectrolytes combine the properties of both electrolytes (salts) and polymers. The following brief discussion of concepts important for the understanding of polyelectrolytes is preceded by an even briefer compilation of important concepts and quantities in the context of electrolyte solutions.

The standard theoretical starting point for the interaction of charges in an electrolyte solution is the **Debye–Hückel theory**. In this theory the potential of a charge q at point $\vec{r}$ from the charge is

$$\phi(\vec{r}) \approx \frac{q}{r} \exp[-r/\lambda_D]. \tag{5.86}$$

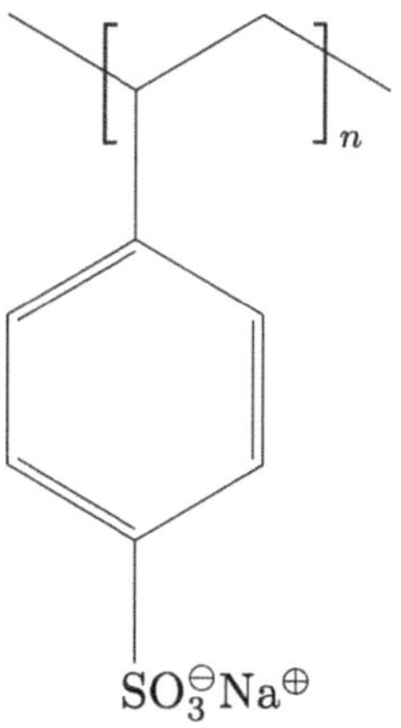

Fig. 5.34 Chemical structure of poly(sodium styrene sulfonate)

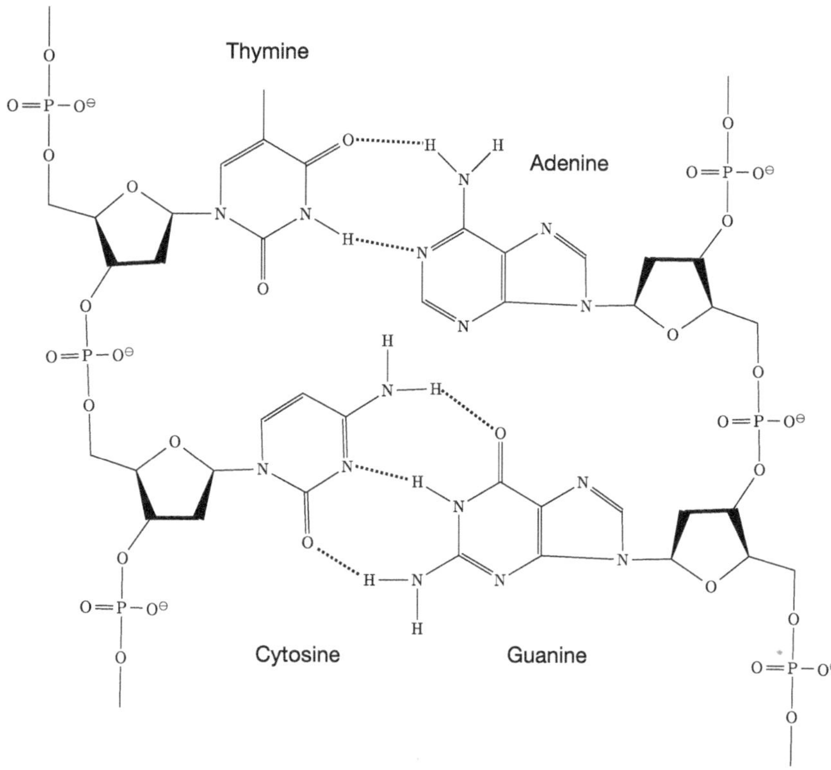

Fig. 5.35 Chemical structure of a section of DNA. Carbon atoms and hydrogen atoms bonded to carbon atoms are not shown explicitly. Note the proton dissociation in the phosphate groups

You can find the discussion of this formula, including the assumptions it involves, in every textbook on physical chemistry and statistical thermodynamics (e.g., [52]). The spatial range of this interaction is determined by the **Debye screening length**

$$\lambda_D = \sqrt{\frac{k_B T}{8\pi e^2 I}}, \tag{5.87}$$

where e is the magnitude of the elementary charge and

$$I = \frac{1}{2}\sum_i c_i z_i. \tag{5.88}$$

is the **ionic strength**. The index i indicates the different types of charges present, c_i is the number concentration of these charges, and z_i is the charging of type i. For example, in the case of NaCl in aqueous solution we have Na^+ and Cl^- ions, i.e. $z_{Na} = +1$ and $z_{Cl} = -1$.

But how large is λ_D? First we note that the system of our current units requires the replacement of q and e by $e/\sqrt{4\pi\varepsilon_o\varepsilon_r}$, if we want to use SI-units. Here ε_r is the dielectric constant of the background medium containing the charges, e.g., ions in water (water: $\varepsilon_r = 78.3$ at $T = 298\,\mathrm{K}$ and $P = 1$ bar). Thus we have

$$\lambda_D = 1.988 \cdot 10^{-3}\sqrt{\frac{T\varepsilon_r}{I}}\ \mathrm{nm} \quad \text{with } [T] = \mathrm{K} \text{ and } [c] = \mathrm{mol/l}. \tag{5.89}$$

For example, in the case of a 0.1 molar aqueous NaCl solution $\lambda_D = 0.96\,\mathrm{nm}$.

A related length is the **Bjerrum length** λ_B defined via

$$\lambda_B = \frac{e^2}{k_B T}. \tag{5.90}$$

λ_B is the length at which the interaction of two charges of magnitude e is equal to the thermal energy $k_B T$. In water at $T = 298\,\mathrm{K}$ we find $\lambda_B \approx 0.7\,\mathrm{nm}$.

A simple model picture of a polyelectrolyte, albeit without additional ions other than the counterions it supplies itself, is that of a (straight) line charge density of e per unit length l surrounded by counterions. The charge density is homogeneously distributed on a cylinder whose radius is r_0. The interaction of a counterion $-e$ at the distance r from the cylinder axis is $-e\Delta\phi_{cyl}(r) = 2e(e/l)\ln(r/r_o)$. Assuming that this is the only relevant interaction in the system and that the counterions are distributed between r_0 and a maximum distance r_1 from the central axis, we obtain the configuration free energy (without the $1/N!$-factor multiplying the partition function) per e and l in units of $k_B T$:

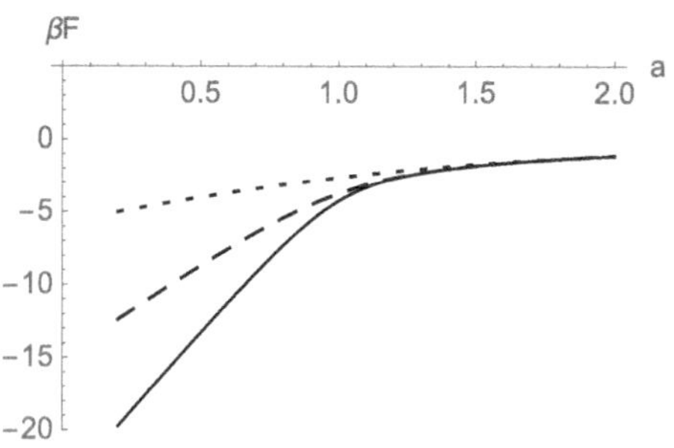

Fig. 5.36 βF versus a for $r_0 = 1$ and $r_1 = 10$ (dotted line), $r_1 = 10^3$ (dashed line), and $r_1 = 10^5$ (solid line)

$$\beta F = -\ln \int_{r_0}^{r_1} 2\pi r \, dr \, \exp[-2a \ln(r/r_o)] \tag{5.91}$$

$$= -\ln \left(\frac{\pi r_0^2 \left(1 - (r_1/r_0)^{2-2a}\right)}{a - 1} \right),$$

where $a \equiv \lambda_B / l$. Figure 5.36 depicts βF versus a for $r_0 = 1$ and $r_1 = 10$ (dotted line), 10^3 (dashed line) and 10^5 (solid line). Obviously, if $a < 1$ or $\lambda_B < l$ the counterions and thereby the rodlike polyelectrolytes favour infinite dilution. For $a > 1$ or $\lambda_B > l$ this is no longer the case. This region is interpreted in terms of a **counterion condensation** occurring (also called **Manning condensation**). Even though this is a very much simplified picture it still provides interesting insight. Nevertheless, we do not want to dwell on this phenomenon. Instead, the interested reader should consult [70, 71].

● *Electrostatic Persistence Length*

It appears natural that the 'stiffness' of a polyelectrolyte should depend on quantities like λ_D or λ_B. This question was addressed first by Skolnick and Fixman [72] and Odijk [73]. Here we employ a simple two-dimensional model for stiff but nevertheless persistent flexible polyelectrolytes (e.g., DNA) to illustrate the main result. Figure 5.37 shows two charges q (in units of e) separated by the distance s along a circular contour. Their direct separation is x. Using $s = \varphi R$ and $(x/2)^2 + R^2 \cos^2(\varphi/2) = R^2$, we obtain for small φ

$$s - x \approx \frac{R\varphi^3}{24}. \tag{5.92}$$

We want to compare the electrostatic interaction energy of the charges in two cases: (i) $R = \infty$, i.e. the circular contour becomes a straight line and the distance

Fig. 5.37 Two charges q separated by a distance s along a circular contour

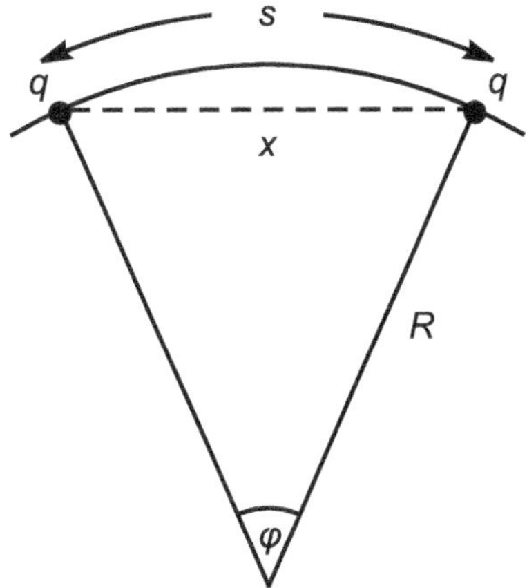

between the charges is s; (ii) R is finite but still large, i.e. the curvature is small so that (5.92) applies, and the distance between the charges is x. The result is

$$q^2 \left(\frac{1}{x} - \frac{1}{s} \right) \approx q^2 \frac{s-x}{s^2} \sim \frac{q^2 s}{R^2}.$$

(5.93)

Next we assume a homogenous line density $1/b$ of charges along the contour screened by mobile countercharges surrounding them. The corresponding total energy difference per charge over the distance of one Debye screening length λ_D is

$$\sim \frac{1}{b} \int_0^{\lambda_D} ds \frac{q^2 s}{R^2} \sim \frac{q^2 \lambda_D^2}{b\, R^2}.$$

(5.94)

Now we compare this result with the bending energy of a persistent flexible polymer in (B.2) of Appendix B. Note that R^{-1} is the curvature and thus $R^{-2} = (\partial \vec{t}/\partial s)^2$. Hence we conclude, based on (B.10) also in Appendix B, that the electrostatic bending stiffness is proportional to λ_D^2 and therefore

$$P_{el} \propto \frac{\lambda_B \lambda_D^2}{b^2},$$

(5.95)

where P_{el} is the **electrostatic persistence length** (Odijk obtains the proportionality constant 1/4). Note that this is merely the electrostatic part of the total persistence length P_{tot}, i.e.

$$P_{tot} = P + P_{el}.$$

(5.96)

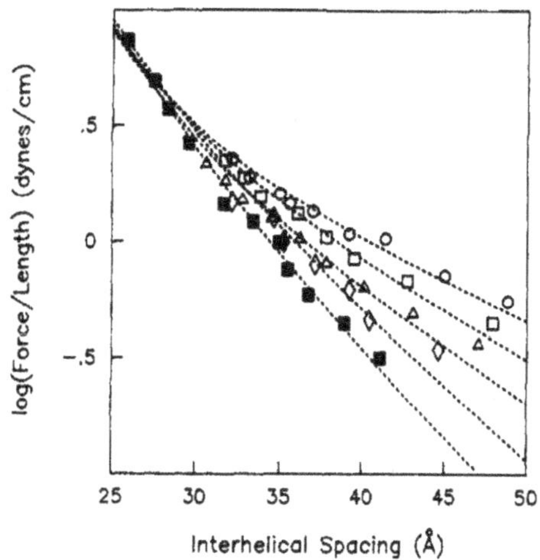

Fig. 5.38 Repulsion between DNA molecules in NaCl solutions. The salt concentrations are 0.2, 0.3, 0.4, 0.8, and 2 M. The lines are fitted according to the theory outlined in the text. Reprinted with permission from [74]. Copyright 1990 American Chemical Society

Here P is the part of the total persistence length contributed by other molecular interactions.

- *Forces Between Persistent Flexible Polyectrolytes in Translationally Ordered Phases*

Shape fluctuations of macromolecules have a profound effect on their interactions at sufficiently elevated concentrations. In the following we discuss the effect of flexibility, i.e. the effect of shape fluctuations, on the forces in hexagonally ordered DNA bundles. We shall discover that molecular flexibility leads to an effective increase of the range of the electrostatic forces.

Figure 5.38 shows experimental data obtained by Podgornik et al. who have measured the repulsive force between hexagonally ordered DNA molecules in NaCl solutions under variable salt concentration as function of the perpendicular packing distance. Basically, the repulsive force decreases exponentially in two distinct regimes. At inter-helical separations less than about 30 Å the salt concentration has no discernible effect. This regime is dominated by the so called **hydration forces**— forces having to do with the structure of water layers surrounding the helix. Hydration forces tend to be stronger than the electrostatic forces controlled by the salt concentration. At larger distances, however, a clearly discernible concentration dependence is observed. Here the forces decay faster when the salt concentration is increased. The question is whether we can construct a theory describing the force between the DNA molecules in this latter regime?

The approach of Podgornik and Parsegian is illustrated in Fig. 5.39. It is assumed that due to the hexagonal packing of the helixes every individual helix is confined to a tube of diameter R, where R is the mean inter-helical distance. Moreover, every segment of length b_{Kuhn} experiences a potential $\phi(\rho)$. In the following we use cylindrical coordinates ρ, φ, and z, where z is measured along the tube axis.

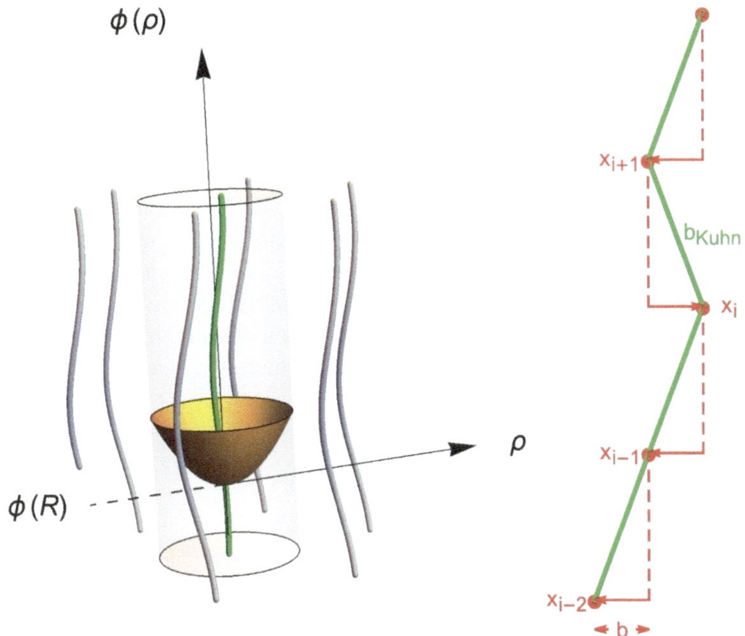

Fig. 5.39 Sketch of a DNA molecule (green) within an effective tube (cylinder) of radius R created by its neighbors (shown in grey) in a hexagonal array of DNA helices. Note that R is the mean separation of the helices. Each DNA segment of length b_{Kuhn} is subject to a harmonic potential $\phi(\rho)$ perpendicular to the cylinder axis illustrated by the paraboloid

Expanding $\phi(\rho)$ around $\rho = R$ yields

$$\phi(\rho) = \phi(R) + \frac{1}{2}\phi''(R)x^2, \tag{5.97}$$

where $x = \rho - R$. We shall not discuss the electrostatic interaction between the DNA helices. Instead we rely on a result obtained by Podgornik and Parsegian, i.e.

$$\phi(R) = \phi_o \frac{e^{-\kappa R}}{\kappa R}. \tag{5.98}$$

Here we assume that κR is large. Hence

$$\phi''(R) \approx \kappa^2 \phi(R). \tag{5.99}$$

In the case at hand the quantity κ is the inverse Debye length. All in all this means that we consider the DNA molecule as being composed of segments of length b_{Kuhn}

experiencing a harmonic potential in the radial direction from the tube's axis. The quantities x_i in Fig. 5.39 indicate the radial position of segments i relative to the minimum of the harmonic potential.

In order to find the free energy of the helices in this model, we employ the **self-consistent field method**. As a matter of fact, it is very convenient to directly use (2.51). One may object that this is a differential equation for one- and not three-dimensional polymers or paths in a potential $u(x)$. However, the changes necessary to upgrade our solution from one to three dimensions are minor and can be made easily at the end. Notice that the radial positions x_i of the segments, which are for the time being just x-positions relative to the potential minimum, are analogous to the arrows in Fig. 2.15 mapping out a conformation of a one-dimensional polymer. This means that we can use (2.51) as it stands and merely replace $u(x)$ by the second term in (5.97), i.e.

$$\left(\frac{b^2}{2} \frac{d^2}{dx^2} + \mu_o - \beta \frac{1}{2} \kappa^2 \phi(R) x^2 \right) \psi_o(x) = 0. \tag{5.100}$$

Here $b = |x_i - x_{i-1}|$, where we assume that b is the same for every i. Now we introduce a new variable s via

$$x = \left(\frac{b^2}{\beta \kappa^2 \phi(R)} \right)^{1/4} s. \tag{5.101}$$

With this substitution we obtain (5.100) in dimensionless form, i.e.

$$\left(\frac{d^2}{ds^2} + w - s^2 \right) \tilde{\psi}_o(s) = 0, \tag{5.102}$$

where

$$\mu_o = \frac{w}{2} b \kappa \left(\beta \phi(R) \right)^{1/2}. \tag{5.103}$$

Equation (5.102) describes the one-dimensional quantum oscillator and for the ground state this means $w = 1$. The total free energy of a segment b_{Kuhn} of the helix is the sum of $\phi(R)$ and $k_B T \mu_o$, i.e.

$$\Delta F(R) = \phi(R) + k_B T \mu_o = \phi(R) + \frac{1}{2} k_B T b \kappa \left(\frac{\phi(R)}{k_B T} \right)^{1/2}. \tag{5.104}$$

This is essentially equation (20) in [74]. The only difference is a numerical (and for the following discussion unimportant) factor in the second term, which arises from the aforementioned reduction of the problem to one-dimension (see below).

But what is the significance of (5.104)? The force shown in Fig. 5.38 is obtained by differentiating $\Delta F(R)$ with respect to R. However, this will not alter the fact that the decay of the first term's contribution to the force is much faster in comparison to the second term, which is contributed by the shape fluctuations of the helix. The square root of $\phi(R)$ in that term essentially doubles the decay length from κ^{-1} to $2\kappa^{-1}$! Of course, changing the salt concentration changes the ionic strength and therefore κ^{-1}. Hence the different slopes of the force data for large inter-helical spacing.

Let's return to the issue of dimension. Podgornik and Parsegian use the three-dimensional version of (5.100). This implies that the DNA-helixes are three-dimensional random walks. The step length l (corresponding to our b) is the same in all three directions. However, it seems to this author that the direction along the tube is special. In particular $b \approx b_{Kuhn}\sqrt{\langle\theta^2\rangle}$, where $\theta(z)$ is the angle between the helix orientation and the tube's axis at point z along the tube. Since the alignment of the helices is quite pronounced, we expect $b \ll b_{Kuhn}$, with b_{Kuhn} comparable to the persistence length of DNA (about 600 Å). The 'upgrade' from one- to two dimensions means that $b^2/2$ in (5.100) is replaced by $b^2/4$ and the ground state value of w is not 1 but 2. Using $b^2/4$ and $w = 2$ does in fact not change (5.104) at all.

The discussion of polyelectrolytes fills books and therefore it is best to stop at this point. However, readers interested in learning more about this important class of polymers may want to start with the review by Muthukumar [75].

5.5 Problems

Problem 5.1 (a) At the beginning of Sect. 5.1 we define **true stress** and **engineering stress**. Previously, in (4.20), we had computed the stress σ_{zz} due to uniaxial stretch of a sample in z-direction at constant volume. Is σ_{zz} in (4.20) the true stress or is it the engineering stress?

(b) What does the answer to (a) imply regarding the statement that (4.21) is a 'reassuring result'? Hint: We had compared (4.19) based on the theory of elasticity with (4.21) based on chain entropy. Since the latter, as we now know, was derived via σ^e_{zz}, we must ask ourselves weather (4.19) is based on σ^e_{zz} as well—and if not, what does this mean with regards to the aforementioned comparison?

Problem 5.2 The internal structure of polymer nano-composite is complex. This means that understanding the relation between internal structure and material properties requires to be able to look inside a sample on different scales. Figure 5.12 in this section contains a number of transmission electron micrographs showing filler distributions inside an elastomer matrix. Therefore TEM offers one way to 'look

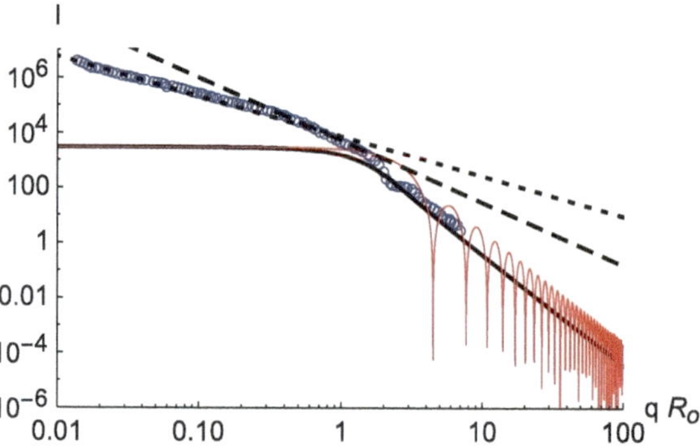

Fig. 5.40 Scattering from particles and particle networks. Open blue circles: SAX intensities I for 9.1% vol. silica particles dispersed in SBR. Red line: Scattering from a homogeneous sphere. Solid line: Average over the intensities from homogeneous spherical particles. Long-dashed line: Fit to the experimental data in an intermediate q-range using $I \propto q^{-d_f}$ with $d_f \approx 2.3$. Short-dashed line: Fit to the experimental data at large q using $I \propto q^{-d_f}$ with $d_f \approx 1.5$. Here $R_o = 44$ Å

inside' a polymer matrix. Another experimental option is X-ray scattering. The symbols in Fig. 5.40 are Small-Angle-X-ray-scattering (SAX) intensities I for 9.1% vol. silica particles dispersed in SBR taken from Fig. 1 (d) in [76]). It is the purpose of this problem to learn what kind of information we may gather from such experiments.

(a)   Calculate the **form factor** of the primary filler particles $f(q)$, assuming that they are perfectly spherical and homogeneous. Use (2.94), where $c(\tau)/m = \rho$ is given by

$$
\rho = \begin{cases} \frac{Q}{4\pi R^3/3} = const & r < R \\ 0 & r > R \end{cases} .
\tag{5.105}
$$

The quantity R is the particle radius, i.e. $\Delta V = 4\pi R^3/3$. Produce a double-log plot of $f^2(q)/Q^2$ versus qR. Note that f^2 is proportional to the scattering intensity.

(b)   Calculate the average of $f^2(q)/Q^2$, using the normalized particle size distribution

$$
g(R) = \frac{16(R/R_o)^6 e^{-(R/R_o)^2}}{15\sqrt{\pi}R_o},
\tag{5.106}
$$

i.e. $\bar{f}^2(q) = \int_0^\infty f^2(q)g(R)dR$. This means that you are looking at the scattering (intensity) from a polydisperse particle population. Show that $\bar{f}^2(q) \propto q^{-4}$ for large q. In the small angle scattering community this is known as **Porod's law** (e.g., [77]). Again produce a log-log-plot of $\bar{f}^2(q)/Q^2$ versus qR_o and compare this graph with the one in part (a) (use $R = R_o$).

(c) Recreate the power-law fits in the above figure and check the values obtained there for d_f. Reminder: We had discussed the scattering from fractal objects in the context of (2.123).

References

1. P.J. Flory, Tensile strength in relation to molecular weight of high polymers. J. Am. Chem. Soc. **67**, 2048 (1945)
2. R.J. Young, P.A. Lovell, *Introduction to Polymers*, 2nd edn. (Chapman & Hall, London, 1991)
3. H.H. Kausch, *Polymer Fracture* (Springer, Heidelberg, 2012)
4. A.J. Kinloch, R.J. Young, *Fracture Behaviour of Polymers* (Springer Science+Business Media, Dordrecht, 1995)
5. M. Mooney, The thermodynamics of a strained elastomer. I. General analysis. J. Appl. Phys. **19**, 434 (1948)
6. R.S. Rivlin, Large elastic deformations of isotropic materials. IV. Further developments of the general theory. Philos. Trans. R. Soc. Lond. Series A Math. Phys. Sci. **241**, 379 (1948)
7. C. Wrana, *Introduction to Polymer Physics* (LanXESS, Leverkusen, 2009)
8. J. Plagge, On the Reinforcement of Rubber by Fillers and Strain-Induced Crystallization. Ph.D. thesis, Universität Hannover (2018)
9. T.A. Vilgis, G. Heinrich, M. Klüppel, *Reinforcement of Polymer Nano-Composites* (Cambridge University Press, Cambridge, 2009)
10. G. Kraus (ed.), *Reinforcement of Elastomers J* (Wiley & Sons, New York, 1965)
11. P. Flory, Thermodynamics of crystallization in high polymers. I. Crystallization induced by stretching. J. Chem. Phys. **15**, 397 (1947)
12. L. Tarrach, A Mesoscopic Computer Model for Reinforcement in Filled and Strain-Crystallizing Elastomers. Ph.D. thesis, University of Wuppertal (2024)
13. S. Trabelsi, P.-A. Albouy, J. Rault, Crystallization and melting processes in vulcanized stretched natural rubber. Macromolecules **36**, 7624 (2003)
14. R. Hentschke, J. Plagge, Strain-induced self-assembly of crystallites in elastomers. Phys. Rev. E **104**, 014502 (2021)
15. J. Plagge, M. Klüppel, A theory relating crystal size, mechanical response, and degree of crystallization in strained natural rubber. Macromolecules **51**, 3711 (2018)
16. P.-A. Albouy, P. Sotta, Draw ratio at the onset of strain-induced crystallization in cross-linked natural rubber. Macromolecules **53**, 992 (2020)
17. J. Plagge, R. Hentschke, Microphase separation in strain-crystallizing rubber. Macromolecules **54**, 5629 (2021)
18. C.G. Robertson, C.J. Lin, M. Rackaitis, C.M. Roland, Influence of particle size and polymer-filler coupling on viscoelastic glass transition of particle-reinforced polymers. Macromolecules **41**, 2727 (2008)
19. M.-J. Wang, M. Morris, *Rubber Reinforcement with Particulate Fillers* (Hanser, Munich, 2021)
20. W.M. Hess, G.C. McDonald, Improved particle size measurements on pigments for rubber. Rubber Chem. Technol. **56**, 892 (1983)
21. L Nikiel, W. Wampler, J. Neilsen, N. Hershberger, How Carbon Black Affects Electrical Properties. Rubber & Plastics News, March 23 (2009)
22. M. Klüppel, The role of disorder in filler reinforcement of elastomers on various length scales. Adv. Polym. Sci. **164**, 1 (2003)
23. H.-D. Luginsland, *A Review on the Chemistry and the Reinforcement of the Silica-Silane Filler System for Rubber Applications* (Shaker Verlag, Aachen, 2002)
24. R. Hentschke, Tensile strength of rubber described via the formation and rupture of load-bearing polymer chains. Phys. Rev. E *106*, 014505 (2022))

25. A. Einstein, Eine neue Bestimmung der Moleküldimensionen. Ann. d. Physik **19**, 289 (1906); Berichtigung zu meiner Arbeit: Eine neue Bestimmung der Moleküldimensionen. Ann. d. Physik **34**, 591 (1911)
26. H.M. Smallwood, Limiting law of the reinforcement of rubber. J. Appl. Phys. **15**, 758 (1944)
27. E. Guth, Theory of filler reinforcement. J. Appl. Phys. **16**, 20 (1945)
28. E. Guth, O. Gold, On the hydrodynamical theory of the viscosity of suspensions. Phys. Rev. **53**, 322 (1938)
29. J. Israelachvili, *Intermolecular & Surface Forces* (Academic, New York, 1991)
30. R. Hentschke, The Payne Effect Revisited. eXPRESS Polymer Lett. **11**, 278 (2017)
31. F. Schön, Elastomer/Schichtsilikat Komposite: Einfluss der Füllstoffstruktur auf mechanische, dynamische und Gasbarriere-Eigenschaften. Ph.D. thesis, Universität Freiburg (2004)
32. P. Meakin, B. Donn, G.W. Mulholland, Collisions between point masses and fractal aggregates. Langmuir **5**, 510 (1989)
33. P. Meakin, Formation of fractal clusters and networks by irreversible diffusion-limited aggregation. Phys. Rev. Lett. **51**, 1119 (1983)
34. M. Kolb, R. Botet, R. Jullien, Scaling of kinetically growing clusters. Phys. Rev. Lett. **51**, 1123 (1983)
35. D.A. Weitz, M. Oliveria, Fractal structures formed by kinetic aggregation of aqueous gold colloids. Phys. Rev. Lett. **52**, 1433 (1984)
36. M. Matsushita, K. Sumida, Y. Sawada, Fractal structure and dynamics for kinetic cluster aggregation of polystyrene colloids. J. Phys. Soc. Jpn. **54**, 2786 (1985)
37. W.D. Brown, R.C. Ball, Computer simulation of chemically limited aggregation. J. Phys. A: Math. Gen. **18**, L517 (1985)
38. P.N. Pusey, J.N. Rarity, Paper presented at Royal Society of Chemistry Symposium on Fractals in Physics and Chemistry, Salford (1986)
39. W. D. Brown, Ph.D. thesis, Department of Physics, University of Cambridge (1987)
40. A.G. Marangoni, L.H. Wesdorp, *Structure and Properties of Fat Crystal Networks*, 2nd edn. (CRC Press, Boca Raton, 2013)
41. M. Klüppel, G. Heinrich, Fractal structures in carbon black reinforced rubbers. Rubber Chem. Technol. **68**, 623 (1995)
42. G. Huber, T.A. Vilgis, Universal properties of filled rubbers: mechanisms for reinforcement on different length scales. Kautschuk Gummi Kunststoffe **52**, 102 (1999)
43. H. Xi, R. Hentschke, The influence of structure on mechanical properties of filler networks via coarse-grained modeling. Macromolecular Theory and Simulation (2014)
44. J. Fritzsche, M. Klüppel, Structural dynamics and interfacial properties of filler-reinforced elastomers. J. Phys.: Condens. Matter *23*, 035104 (2011)
45. M. Loos, *Carbon Nanotube Reinforced Composites: CNR Polymer Science and Technology.* (Elsevier Inc, 2015)
46. A.N. Semenov, A.R. Khokhlov, Statistical physics of liquid-crystalline polymers. Sov. Phys. Usp. **31**, 988 (1988)
47. R. Hentschke, Equation of state for persistent-flexible liquid-crystal polymers. Comparison with Poly(γ-benzyl-L-glutamate) in Dimethylformamide. Macromolecules **23**, 1192 (1990)
48. L. Onsager, The effects of shape on the interaction of colloidal particles. Ann. N.Y. Acad. Sci. *51*, 988 (1949)
49. M. Abramowitz, I.A. Stegun (eds.), *Handbook of Mathematical Functions* (Dover, New York, 1972)
50. R. Hentschke, B. Fodi, Theory of the supramolecular liquid crystal, in *Supramolecular Polymers*, ed. by A. Ciferri, 1st edn. (Marcel Dekker, New York, 2000)
51. R. Hentschke, J. Herzfeld, Isotropic, nematic, and columnar ordering in systems of persistent flexible hard rods. Phys. Rev. A **44**, 1148 (1991)
52. R. Hentschke, *Thermodynamics*, 2nd edn. (Springer, Heidelberg, 2022)
53. J.O. Hirschfelder, C.F. Curtiss, R.B. Bird, *Molecular Theory of Gases and Liquids* (Wiley, New York, 1954)
54. T. Odijk, Theory of lyotropic polymer liquid crystals. Macromolecules **19**, 2313 (1986)

55. G.J. Vroege, H.N.W. Lekkerkerker, Phase transitions in lyotropic colloidal and polymer liquid crystals. Rep. Prog. Phys. **55**, 1241 (1992)
56. H.N.W. Lekkerkerker, G.J. Vroege, Lyotropic colloidal and macromolecular liquid crystals. Philos. Trans.: Phys. Sci. Eng. **344**, 419 (1993)
57. A.M. Donald, A.H. Windle, S. Hanna, *Liquid Crystalline Polymers* (Cambridge University Press, Cambridge, 2005)
58. K. Binder, S.A. Egorov, A. Milchev, A. Nikoubashman, Understanding the properties of liquid-crystalline polymers by computational modeling. J. Phys. Mater. **3**, 032008 (2020)
59. I.M. Ward (ed.), *Structure and Properties of Oriented Polymers* (Springer, Dordrecht, 1997)
60. D.F. Evans, H. Wennerstrom, *The Colloidal Domain* (VCH Publishers, Weinheim, 1994)
61. M.P. Taylor, J. Herzfeld, Liquid-crystal phases of self-assembled molecular aggregates. J. Phys.: Condens. Matter **5**, 2651 (1993)
62. W.M. Gelbart, A. Ben-Shaul, D. Roux (eds.), *Micelles, Membranes, Microemulsions, and Monolayers* (Springer, Heidelberg, 1994)
63. B. Widom, K.A. Dawson, M. D. Lipkin, in *Statistical Physics: Invited lectures from STATPHYS 16*, ed. by H.E. Stanley (North-Holland, Amsterdam, 1986)
64. T. Odijk, Effect of micellar flexibility on the isotropic-nematic phase-transition in solutions of linear aggregates. J. Physique **48**, 125 (1987)
65. R. Hentschke, Effect of persistent-flexibility on the isotropic, nematic, and columnar ordering in a self-assembling system. Liquid Cryst. **10**, 691 (1991)
66. P. van der Schoot, M.E. Cates, The isotropic-to-nematic transition in semi-flexible micellar solutions. Europhys. Lett. **25**, 515 (1994)
67. R. Hentschke, J. Herzfeld, Soft repulsions in a lattice model for self-assembling systems. J. Chem. Phys. **91**, 7308 (1989)
68. R.E. Dickerson, I. Geis, *Hemoglobin: Structure, Function, Evolution and Pathology* (Benjamin/Cummings, Amsterdam, 1983)
69. R. Hentschke, J. Herzfeld, Theory of nematic order with aggregate dehydration for reversibly assembling proteins in concentrated solutions: application to sickle-cell hemoglobin polymers. Phys. Rev. A **43**, 7019 (1991)
70. G.S. Manning, The critical onset of counterion condensation: a survey of its experimental and theoretical basis. Ber. Bunsen. phys. Chem. **100**, 909 (1996)
71. G.S. Manning, Counterion condensation theory of attraction between like charges in the absence of multivalent counterions. Eur. Phys. J. E **34**, 132 (2011)
72. J. Skolnick, M. Fixman, Electrostatic persistence length of a wormlike polyelectrolyte. Macromolecules **10**, 944 (1977)
73. T. Odijk, Polyelectrolytes near the Rod Limit. J. Poly. Sci.: Polymer Phys. *15*, 477 (1977)
74. R. Podgornik, V.A. Parsegian, Molecular fluctuations in the packing of polymeric liquid crystals. Macromolecules **23**, 2265 (1990)
75. M. Muthukumar, 50th anniversary perspective: a perspective on polyelectrolyte solutions. Macromolecules **50**, 9528 (2017)
76. A. Bouty, L. Petitjean, C. Degrandcourt, J. Gummel, P. Kwaśniewski, F. Meneau, F. Boué, M. Couty, J. Jestin, Nanofiller structure and reinforcement in model silica/rubber composites: a quantitative correlation driven by interfacial agents. Macromolecules **47**, 5365 (2014)
77. O. Glatter, O. Kratky (eds.), *Small Angle X-ray Scattering* (Academic, New York, 1982)

Appendix A
Phenomenological Models for Viscoelasticity

Abstract The appendix contains a continuation of our discussion of phenomenological models for viscoelasticity and persistent flexibility. In addition, suggestions are made on how to teach and study the material in this book. Finally, there are hints and solutions to the problems.

The **Kelvin-Voigt model** is only one of several simple combinations of the basic elements and, in addition, is not applicable in the full frequency range. Our next model is the so called **Maxwell model** (model (b)) in Fig. 4.8. This model arranges the two basic elements in series. Its mathematical description is

$$\sigma = \sigma_G = \sigma_\eta \qquad \gamma = \gamma_G + \gamma_\eta \ . \tag{A.1}$$

Inserting (4.50) and (4.51) we obtain

$$(b): \qquad \sigma + \frac{\eta}{G}\dot{\sigma} = \eta\dot{\gamma} \ . \tag{A.2}$$

The same ansatz as before leads to

$$G'/G = \frac{\tau_M^2\omega^2}{\tau_M^2\omega^2 + 1} \tag{A.3}$$

$$G''/G = \frac{\tau_M\omega}{\tau_M^2\omega^2 + 1} \tag{A.4}$$

$$\tan\delta = \frac{1}{\tau_M\omega} \ , \tag{A.5}$$

where

$$\tau_M = \eta/G \ . \tag{A.6}$$

R. Hentschke, *A Concise Introduction to Polymer Physics,*, Undergraduate Lecture Notes in Physics, https://doi.org/10.1007/978-3-031-87324-9

Yet another phenomenological model is the **Zener model** (model (c) in Fig. 4.8). The mathematical description of the Zener model is slightly more complex compared to the Kelvin-Voigt or the Maxwell model. Now we have

$$\sigma = \sigma_1 + \sigma_M$$
$$\sigma_M = \sigma_2 = \sigma_\eta$$
$$\gamma = \gamma_1 = \gamma_M$$
$$\gamma_M = \gamma_2 + \gamma_\eta \, .$$

The indices 1 and 2 refer to G_1 and G_2. Again we employ (4.50) and (4.51), which here leads to

$$(c) : \sigma + \frac{\eta}{G_2}\dot\sigma = G_1\gamma + \eta\left(1 + \frac{G_1}{G_2}\right)\dot\gamma \, . \tag{A.7}$$

As before the relation between stress and strain is linear and the above ansatz (real or complex) yields

$$G'/G_1 = \frac{\tau_2^2\omega^2/\theta + 1}{\tau_2^2\omega^2 + 1} \tag{A.8}$$

$$G''/G_2 = \frac{\tau_2\omega}{\tau_2^2\omega^2 + 1} \tag{A.9}$$

$$\tan\delta = \frac{1-\theta}{\theta}\frac{\tau_2\omega}{\tau_2^2\omega^2/\theta + 1} \, , \tag{A.10}$$

where $\tau_2 = \eta/G_2$ and $\theta = G_1/(G_1 + G_2)$. The quantity τ_2 again describes a relaxation time. If the instantaneous strain $\gamma(t = 0) = \gamma_o$ is held constant, i.e. $\dot\gamma(t) = 0$ for $t > 0$, then

$$\sigma(t) = \left(G_1 + G_2 e^{-t/\tau_2}\right)\gamma_o \, . \tag{A.11}$$

Question: What is the justification for $\sigma(t = 0) = (G_1 + G_2)\gamma_o$? (The strain occurs instantaneously at $t = 0$. The dashpot cannot follow as quickly, i.e. only the G-elements contribute to the answer.).

We briefly want to note that the maximum of $\tan\delta$ in the Zener model is at

$$\tau_2\omega_{max} = \sqrt\theta \tag{A.12}$$

and the attendant value of $\tan\delta(\omega_{max})$ is

$$\tan\delta|_{max} = \frac{1-\theta}{2\sqrt\theta} = \frac{1 - (\tau_2\omega_{max})^2}{2\tau_2\omega_{max}} \, . \tag{A.13}$$

Let us study the low and the high frequency limits of the (A.8) through (A.10). The limit of small ω leads to the Kelvin-Voigt model (with $G_1 = G$). In the opposite limit of large ω we obtain $G' \approx G_1 + G_2$, $G'' \approx G_2/(\tau_2\omega)$, and $\tan\delta \approx (G_2/(G_1 + G_2))/(\tau_2\omega)$. The special case $G_1 \ll G_2$ yields the Maxwell model in the same limit (with $G_2 = G$). Figure A.1 shows the various results of the Zener model for $G_1 = G_2 = G$. The dashed lines are the leading contributions in the above two limits. In order to assess the value of our simple models we must relate their results to actual measurements.

Figure A.2 shows such measurements in comparison to the Zener model. Notice that the values for G_1 and G_2 follow by fitting the theoretical storage modulus to the experimental results in the respective limits at low and high frequencies. The value of $\tau_2 \approx 10^{-7}$ s is obtained by adjusting the inflection point of the theoretical storage modulus to the data. Despite its simplicity the Zener model provides an overall qualitatively correct description of the data.

Before we explore possible improvements of the Zener model, we should make sure that we do understand why it works the way it works. After all, the model combines three simple basic elements, i.e. two springs plus a dashpot, into a reasonable description of the mechanical properties of a complex system.

Figure A.3 shows a number of alternative combinations of the basic models, i.e. (a) (Maxwell: M) and (b) (Kelvin-Voigt: KV) depicted in Fig. 4.8. If we investigate the mechanical behavior of the new models, then the upper row behaves very much like the Maxwell model, i.e. $\tan\delta \sim \omega^{-1}$, whereas the lower row behaves according to the Kelvin-Voigt model, i.e. $\tan\delta \sim \omega$. None of the models reproduces the maximum of $\tan\delta$ versus ω. Only the Zener model (sometimes also called Poynting-Thomson relaxation model) and its dual partner, both depicted in Fig. A.4, do yield qualitative agreement with the experiment (The two models can be converted into one another via $G_1^R = G_1^K G_2^K/(G_1^K + G_2^K)$, $G_2^R = G_2^{K^2}/(G_1^K + G_2^K)$ and $\eta^R = (G_2^K/(G_1^K + G_2^K))^2 \eta^K$. The index R indicates the relaxation or Zener model, the index K indicates its dual partner.). Notice that interchanging springs and dash pots in the two models does not yield useful results or insights.

We can understand our above results as follows. At low frequencies an increase of $\tan\delta$ is observed. This means that the friction element, i.e. the dashpot, must be able to follow the excitation. This is build into the Kelvin-Voigt model, because the amplitudes in the two branches are strictly coupled. The decrease of $\tan\delta$ at high frequencies, on the other hand, requires the decoupling of the friction elements from the excitation. This is build into the Maxwell model, where the spring can take over the strain from the friction element. The Zener model incorporates both behaviors.

Remark: A parameter we have mentioned is the strain amplitude. There is no amplitude dependence of the dynamic moduli predicted by the above models. This is because they are linear. Most elastomer materials contain large amounts of filler. In the case of automobile tires the fillers are carbon black and/or silica nanoparticles. The resulting materials are highly non-linear and their dynamic moduli are strongly dependent on the strain amplitude (**Payne effect**).

Fig. A.1 Results of the
Zener model

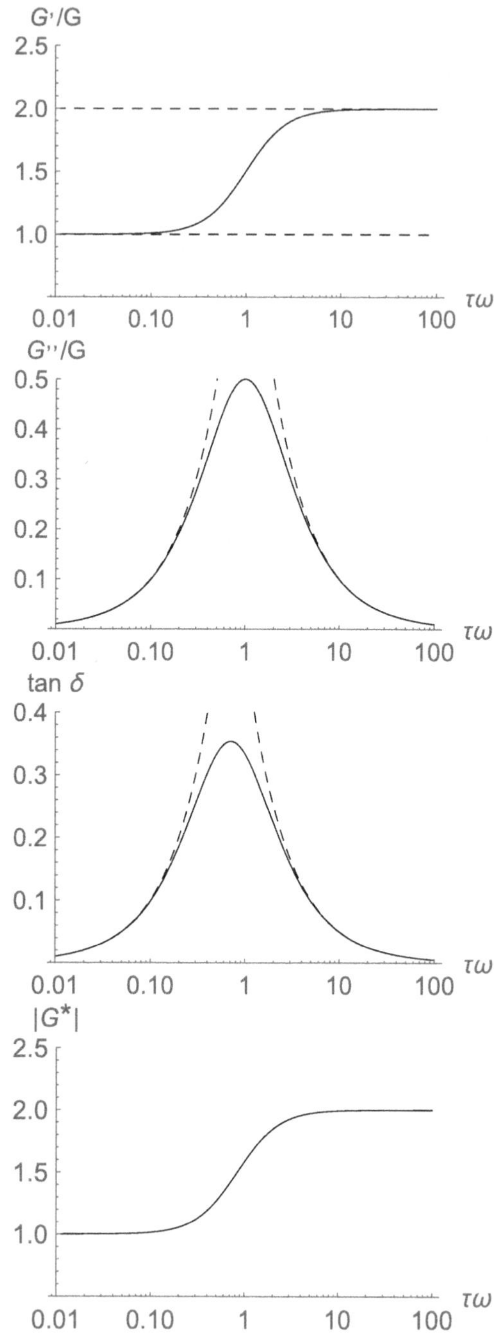

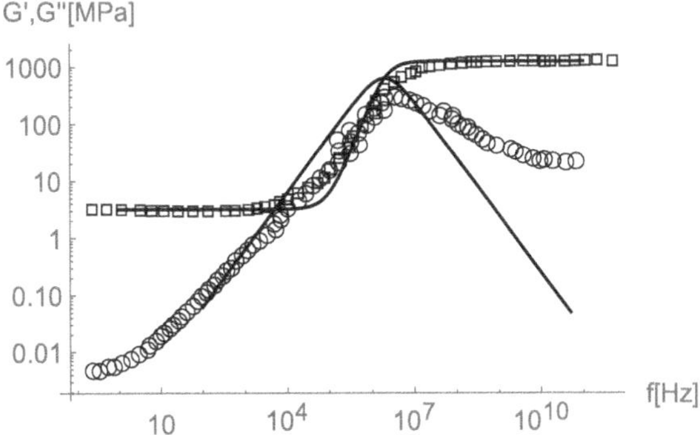

Fig. A.2 Dynamic moduli of the Zener model (lines) in comparison to measured data (squares: storage modulus; circles: loss modulus) obtained for a highly cross-linked polyisoprene rubber versus strain frequency (data reproduced with the permission of Continental Reifen Deutschland). Note that the loss modulus data in this figure are the same as in Fig. 4.32. Here the values for G_1 and G_2 are different from the values used in Fig. 4.32 in order to obtain better overall agreement with both moduli

Fig. A.3 Simple combination models

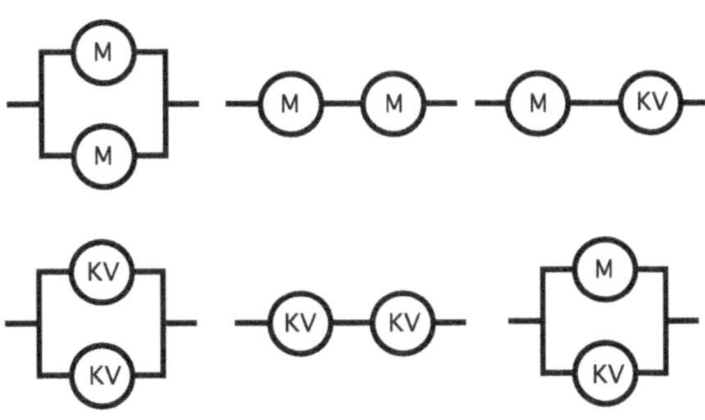

Fig. A.4 Top: Zener or Poynting-Thomson relaxation model; bottom: dual partner

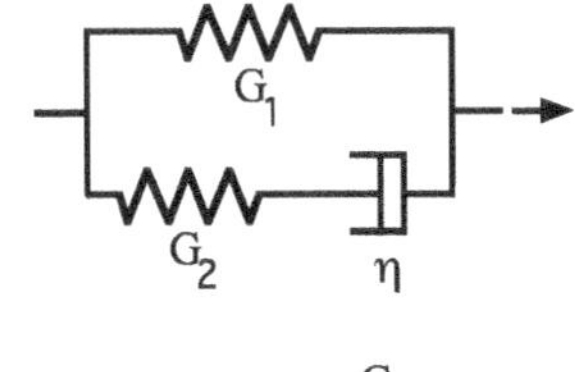

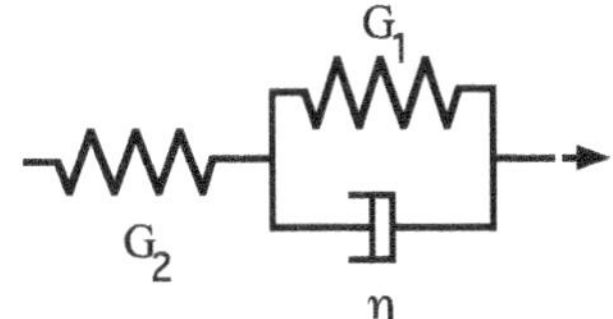

Appendix B
Persistence Length from Fluctuation Theory

Note that H in (5.36) can be expressed as an integral (continuum limit). To see this note that

$$(\vec{t}_i - \vec{t}_{i+1})^2 = 2 - 2\,\vec{t}_i \cdot \vec{t}_{i+1} \ . \tag{B.1}$$

Hence

$$H = H_o + \frac{b^2 J}{2} \sum_{i=1}^{N} \left(\frac{\vec{t}_i - \vec{t}_{i+1}}{b} \right)^2 \stackrel{c.l.}{=} H_o + \frac{1}{2} bJ \int_0^L ds \left(\frac{\partial \vec{t}}{\partial s} \right)^2 , \tag{B.2}$$

where $H_o = -NJ$. *c.l.* means $N \to \infty$, $b \to 0$ while $L = Nb$. Note that bJ is a bending stiffness. The probability of a certain contour conformation is

$$p(\{\vec{t}\}) \propto \exp\left[-\beta \int_0^L ds \left\{ \frac{\lambda}{L} \vec{t}^{\,2}(s) + \frac{1}{2} bJ \left(\frac{\partial \vec{t}}{\partial s} \right)^2 \right\} \right] \tag{B.3}$$

$(\beta = (k_B T)^{-1})$. Here λ is a **Lagrange multiplier** due to the condition

$$\frac{1}{L} \int_0^L ds \, \vec{t}^{\,2}(s) = 1 , \tag{B.4}$$

which is the global version of the local condition $\vec{t}^{\,2}(s) = 1$. According to a general result of fluctuation theory (e.g., [1]; p. 228 ff)

$$\langle \hat{t}_q \cdot \hat{t}_{-q} \rangle = \frac{d-1}{2\beta} \frac{1}{\lambda + \frac{L}{2} bJ \, q^2} , \tag{B.5}$$

where $\langle \hat{t}_q \cdot \hat{t}_{-q} \rangle$ is the Fourier transform of $\langle \vec{t}(0) \cdot \vec{t}(s) \rangle$. Note that $d-1$ distinguishes two cases: $d = 3$ is a polymer in three-dimensional space and $d = 2$ is a polymer in

R. Hentschke, *A Concise Introduction to Polymer Physics,*, Undergraduate Lecture Notes in Physics, https://doi.org/10.1007/978-3-031-87324-9

two-dimensional space. In the former case there are two fluctuating angles underlying the local orientation of $\vec{t}$, whereas in the latter case there is only one (for details see §127 in [2]). Hence

$$\langle \vec{t}(0) \cdot \vec{t}(s) \rangle = \frac{L}{2\pi} \int_{-\infty}^{\infty} \langle \hat{t}_q \cdot \hat{t}_{-q} \rangle e^{iqs} dq = \frac{d-1}{4\beta} \left(\frac{2L}{\lambda b J} \right)^{1/2} e^{-s/P} , \quad (B.6)$$

where

$$P = \left(\frac{LbJ}{2\lambda} \right)^{1/2} . \quad (B.7)$$

The Lagrange multiplier follows via $\langle \vec{t}^2(0) \rangle = 1$, i.e.

$$\lambda = \frac{L(d-1)^2}{8\beta^2 b J} , \quad (B.8)$$

and the final result becomes

$$\langle \vec{t}(0) \cdot \vec{t}(s) \rangle = e^{-s/P} \quad (B.9)$$

with

$$P = \frac{2}{d-1} \frac{bJ}{k_B T} . \quad (B.10)$$

For $d = 3$ this agrees with (2.22) and (5.39), respectively!

References

1. R. Hentschke, *Statistische Mechanik* (Wiley-VCH, Weinheim, 2004)
2. L.D. Landau, E.M. Lifshitz, *Statistical Mechanics* (Butterworth-Heinemann, Amsterdam, 1987)

Appendix C
Teaching and Studying the Material in these Notes

The following table shows the material in these notes organized into 24 lectures (1.5 h each). This should be useful for potential lecturers. In addition, students, who wish to study the material by themselves, may benefit from the partitioning of the course material into more easily digestible increments.

Lecture	Content	Pages
	POLYMER MICROSTRUCTURE, CLASSIFICATION, AND MASS	
L 1	Molecular Microstructure and Classification, Molecular Mass	1–9
L 2	Flexibility Mechanisms and Polymer Dimension	11–21
	EQUILIBRIUM CONFORMATION OF SINGLE CHAINS	
L 3	RIS/Transfer Matrix Approach	22–30
L 4	Self-Consistent Field Approach	30–33
L 5	Conformation Entropy, Flory's Exponent	33–38
L 6	The Scaling Concept	38–44
L 7	Scattering from Ideal Chains	44–50
L 8	Light Scattering and Zimm Plot	51–55
L 9	Scattering from Real Chains	55–58
	THERMODYNAMICS OF BLENDS, SOLUTIONS, AND NETWORKS	
L 10	Lattice Model for Binary Polymer Mixtures	63–67
L 11	Phase Separation in Polymer Mixtures and Polymers in Solution	67–73
L 12	Swelling of Polymer Networks	74–81
	POLYMER DYNAMICS	

© The Editor(s) (if applicable) and The Author(s),under exclusive license to Springer Nature Switzerland AG 2025

R. Hentschke, *A Concise Introduction to Polymer Physics,*, Undergraduate Lecture Notes in Physics, https://doi.org/10.1007/978-3-031-87324-9

Lecture	Content	Pages
L 13	Linear Deformation Mechanics, Ideal Fluids, Phenomenological Models of the Dynamic Moduli	85–105*
L 14	Time-Temperature Superposition, Shear Relaxation Modulus and Relation to Storage and Loss Modulus	105–111
L 15	Single Chain Dynamics: Preliminaries (friction, Brownian motion, equation of motion of the bead-spring chain with friction)	112–116**
L 16	Single Chain Dynamics: Rouse Model—Solution and Results	116–121
L 17	Single Chain Dynamics: Zimm Model—Hydrodynamic Interactions and Their Effects	121–130***
L 18	Polymer Entanglement	130–134
L 19	The Glass Process	134–146
	SELECTED TOPICS	
L 20	Aspects of Polymer Mechanics	151–162
L 21	Filler Effects	163–178
L 22	Liquid Crystalline Polymers	178–191
L 23	Reversibly Assembling Polymers	191–197
L 24	Polyelectrolytes	197–205

(*) : This lecture focusses on the straightforward introduction of (4.13) and (4.49) and how they form the building blocks of the phenomenological models of linear viscoelasticity. This means that most of the content of the two boxes *Equations and Concepts from Isotropic Elasticity* and *Equations and Concepts from Fluid Mechanics* is omitted and left as a reading assignment (or perhaps an additional lecture). Based on (4.13) and (4.49), the Kelvin-Voigt model is discussed in relation to the damped harmonic oscillator under the influence of an external periodic force, since the latter is studied extensively in undergraduate physics. In addition, the storage and the loss modulus are introduced.

(**): including (4.107)

(***): There are a number of detailed calculations on these pages. For the better part they should be omitted in class. Instead one should focus on the main results and the differences between the models of Rouse and Zimm.

Appendix D
Solutions to the Problems

- *Problem* 1.1 After multiplication of the equation by $\bar{M}_n$ the left hand side becomes

$$\left(\bar{M}_w - \bar{M}_n\right)\bar{M}_n = \frac{\sum_i N_i M_i^2}{\sum_j N_j M_j}\frac{\sum_k N_k M_k}{\sum_l N_l} - (\bar{M}_n)^2 = \bar{M}_n^2 - (\bar{M}_n)^2 . \quad \text{(D.1)}$$

- *Problem* 1.2 We can try to extract the first few data points along each of the curves and extrapolate them to zero concentration. We can also just take a curved ruler to do the same by hand (where 0.1 on the log-scale serves as zero). The overall accuracy is roughly the same. Using the second method I obtain $\Pi/c \cdot 10^{-2}|_{c\to 0}$ [cm] $\approx 3.6, 1.3$, and 0.5. Note that in SI-units $[\Pi/c] = (\text{m/s})^2$. Hence we must multiply the Π/c-data by $g = 9.81$ m/s^2 and we should not forget to convert from centimetre to meter. With this we obtain $\bar{M}_n \approx 7 \cdot 10^4$ g/mol, $19 \cdot 10^4$ g/mol, and $51 \cdot 10^4$ g/mol. These numbers are very close to the results in the paper obtained for the samples denoted α-104, α-12, and α-103.

Remark: For the sake of comparison—the molecular weight of water is 18 g/mol.

- *Problem* 2.1 (a) We can simply check this formula by multiplying both sides with $(\mathbf{I} - \langle \mathbf{t}\rangle)^2$. Note that $1/(\mathbf{I} - \langle \mathbf{t}\rangle)$ is the inverse of $\mathbf{I} - \langle \mathbf{t}\rangle$. On the right hand side we obtain $(N-1)\langle \mathbf{t}\rangle - N\langle \mathbf{t}\rangle^2 + \langle \mathbf{t}\rangle^{N+1}$. On the left hand side we find

$$
\begin{aligned}
&(N-1)(\langle \mathbf{t}\rangle \; -2\langle \mathbf{t}\rangle^2 \; +\langle \mathbf{t}\rangle^3) \\
&+(N-2)(\quad\;\; \langle \mathbf{t}\rangle^2 \quad -2\langle \mathbf{t}\rangle^3 \; +\langle \mathbf{t}\rangle^4) \\
&+(N-3)(\qquad\qquad \langle \mathbf{t}\rangle^3 \quad -2\langle \mathbf{t}\rangle^4 \; +\langle \mathbf{t}\rangle^5) \\
&+(N-4)(\qquad\qquad\qquad\; \langle \mathbf{t}\rangle^4 \quad -2\langle \mathbf{t}\rangle^5 \; +\langle \mathbf{t}\rangle^6) \\
&+(N-..)(\qquad\qquad\qquad\qquad\qquad \cdots \qquad \cdots \\
&\qquad\quad \cdots
\end{aligned}
$$

Looking at this schema we see that only the first, the second and the last column yield non-zero terms and these terms are identical to what we found before. If

© The Editor(s) (if applicable) and The Author(s),under exclusive license to Springer
Nature Switzerland AG 2025
R. Hentschke, *A Concise Introduction to Polymer Physics,*, Undergraduate Lecture Notes
in Physics, https://doi.org/10.1007/978-3-031-87324-9

you are uncomfortable with this 'reverse engineering' approach, you may work out $\sum_{i>j} q^{i-j}$, where $q < 1$ is a positive number. This is a simple exercise in geometric series and the result will have the same form as the right hand side of the equation in the original problem. Then you can replace q by $\langle t \rangle$ and proceed as before.

(b) The formula is

$$C_N = (\mathbf{I} + \langle \mathbf{t} \rangle) \frac{1}{\mathbf{I} - \langle \mathbf{t} \rangle}\Big|_{11} - \frac{2}{N} \langle \mathbf{t} \rangle \left(\mathbf{I} - \langle \mathbf{t} \rangle^N\right) \left(\frac{1}{\mathbf{I} - \langle \mathbf{t} \rangle}\right)^2 \Big|_{11}. \tag{D.2}$$

(c) Note that $\langle R^2 \rangle$ is given by (2.8), where $\nu = 1/2$, and the contour length is $L = bN$ (cf. (2.1) with $\phi \approx \pi$). The matrix $\langle \mathbf{t} \rangle$ is given in (2.23) and (2.25) relates the persistence length to $\langle \theta^2 \rangle$. A final piece of information is

$$\left(1 - \frac{\langle \theta^2 \rangle}{2}\right)^N \approx e^{-L/P}, \tag{D.3}$$

where we make use of (2.25) and $L = bN$ in conjunction with $(1 - x/m)^m \approx \exp[-x]$ for large m. With this information it is straightforward to work out the right hand side of the equation for C_N. To obtain the desired formula we also make use of $b \ll P$.

(d) Expansion of $\langle R^2 \rangle$ in the limits $L \gg P$ and $L \ll P$ yields

$$\langle R^2 \rangle \approx \begin{cases} 2L\,P\left(1 - \frac{P}{L}\right) & (L \gg P) \\ L^2\left(1 - \frac{1}{3}\frac{L}{P}\right) & (L \ll P) \end{cases}. \tag{D.4}$$

Hence, (2.27) is the leading behavior of $\langle R^2 \rangle$ in the limit $L \gg P$.

- *Problem 2.2* (a) The transfer matrix is given by

$$\mathbf{T} = \begin{pmatrix} e^{-J/T} & e^{J/T} \\ e^{J/T} & e^{-J/T} \end{pmatrix}. \tag{D.5}$$

We obtain the eigenvalues λ_1 and λ_2 by calculating the determinant

$$det \begin{pmatrix} x^{-1} - \lambda & x \\ x & x^{-1} - \lambda \end{pmatrix}, \tag{D.6}$$

where $x = e^{J/T}$. Hence $\lambda_1 = (1 + x^2)/x$ and $\lambda_2 = (1 - x^2)/x$.

(b) The internal energy E is given by $E = -\partial_\beta \ln Q$. With $Q = \lambda_1^N + \lambda_2^N \to ((1 + x^2)/x)^N$ for large N. Thus $E \to Nk_B J(1 - x^2)/(1 + x^2)$. The heat capacity $C = \partial E/\partial T$ becomes $C \to Nk_B (J/T)^2 (2x/(1 + x^2))^2$. The probability $p_T(\vartheta = 1)$ is computed via

$$p_T(\vartheta = 1) = Q^{-1} \mathrm{Tr} \left(\begin{pmatrix} 1 & 0 \\ 0 & 0 \end{pmatrix} \mathbf{T} \right) \tag{D.7}$$

Using the diagonalization procedure discussed in the text, where in this case

$$S = \begin{pmatrix} -1 & 1 \\ 1 & 1 \end{pmatrix} , \tag{D.8}$$

we obtain $p_T(\vartheta = 1) = 1/2$—the obvious result.

• *Problem* 2.3 (a) The total partition function is the product of n identical single-segment partition functions q. Each segment can be either straight or bent (left or right). Hence

$$Q = q^n = \left(e^{-\beta\epsilon_1} + 2e^{-\beta\epsilon_2}\right)^n \tag{D.9}$$

($\beta = 1/(k_B T)$). Using $\epsilon_1 = 0$ and $\epsilon_2 = \epsilon$ this becomes

$$Q = \left(1 + 2e^{-\beta\epsilon}\right)^n . \tag{D.10}$$

(b) The internal energy of the chain is given by

$$E = \frac{\partial}{\partial(-\beta)} \ln Q = n \frac{2\epsilon e^{-\beta\epsilon}}{1 + 2e^{-\beta\epsilon}} . \tag{D.11}$$

(c) In the limit of infinite temperature $\beta\epsilon \to 0$ and thus

$$Q \to 3^n \quad \text{and} \quad E \to \frac{2}{3}n\epsilon . \tag{D.12}$$

We find the entropy via $F = E - TS$, i.e.

$$S = \frac{E - F}{T} \to -\frac{F}{T} = k_B \ln Q = k_B n \ln 3 . \tag{D.13}$$

Of course, this can be found much faster via $S = k_B \ln \Omega$, where Ω is the number of equivalent microstates, i.e. $\Omega = 3^n$ in this case.

Problem 2.4 (a)

$$\frac{\Delta S}{k_B} = \frac{\Delta E}{k_B T} - \frac{\Delta F}{k_B T}$$

$$\stackrel{(2.50)}{=} \frac{n}{k_B T} \int dx\, c(x) u(x) - n\mu_0$$

$$= n \int dx\, c(x) \left[\beta u(x) - \mu_0\right] \tag{D.14}$$

$$\stackrel{(2.51)}{=} n \int dx\, \frac{a^2}{2} \psi_0(x) \frac{d^2}{dx^2} \psi_0(x)$$

$$\overset{\text{p.\,i.}}{=} -n\frac{a^2}{2}\int dx\left(\frac{d\psi_0(x)}{dx}\right)^2 .$$

Here p. i. stands for 'partial integration'. Note that $\psi_0(x)$ vanishes for $x = \pm\infty$. Hence

$$\frac{\Delta F}{nk_BT} = \frac{\Delta E}{nk_BT} + \int dx \frac{a^2}{2}\left(\frac{d\psi_0}{dx}\right)^2 . \tag{D.15}$$

With $n^{-1}\Delta E = \int dx\, c(x)u(x)$ we obtain the desired relation.

(b) $\psi_0(x)$ follows from (2.45) via the variation

$$0 = \frac{\delta}{\delta\psi_0(x)}\left(\frac{\Delta F}{nk_BT} + \lambda\left[\int dx\,\psi_0^2(x) - 1\right]\right) , \tag{D.16}$$

where λ is a Lagrange multiplier, i.e.

$$0 = \int dx\left[2\frac{a^2}{2}\frac{d\psi_0(x)}{dx}\frac{d\delta\psi_0(x)}{dx} + 2\beta u(x)\psi_0(x)\delta\psi_0(x)\right.$$
$$\left. + 2\lambda\psi_0(x)\delta\psi_0(x)\right] \tag{D.17}$$
$$\overset{\text{p.\,I.}}{=} \int dx\left[-\frac{a^2}{2}\frac{d^2}{dx^2}\psi_0(x) + \left(\beta u(x) + \lambda\right)\psi_0(x)\right]\delta\psi_0(x) .$$

For arbitrary $\delta\psi_0(x)$ this implies

$$\frac{a^2}{2}\frac{d^2}{dx^2}\psi_0(x) + \left(-\lambda - \beta u(x)\right)\psi_0(x) = 0 . \tag{D.18}$$

And setting $-\lambda = \mu_0$ we again obtain (2.51).

• *Problem* 2.5 First we compute $\langle s\rangle = 1/2$ and $\sigma_s = 1/\sqrt{12}$. Then we generate S_n-histograms based on 10^5 S_n-values each. The results are depicted in the two graphs in Fig. D.1. Note that n does not have to be very large in order to generate a standard normal distribution to good approximation.

• *Problem* 2.6 The answer is that it is hidden in (2.49). The right hand side of (2.49) is the first term in a sum over terms $\exp[-n\mu_i\psi_i(x')\psi_i(x)]$, where $i = 0, 1, \ldots$ indicates the eigenvalues and eigenfunctions. Ground state dominance, i.e. the first term is sufficient, here implies $n(\mu_1 - \mu_0) \gg 1$. If we compare (2.48) with Schrödinger's equation for the infinite potential well of width D (see any book on introductory quantum mechanics) we find that the eigenvalues μ_i are given by $\mu_i \sim (b/D)^2(i+1)^2$. Hence, $n(\mu_1 - \mu_0) \gg 1$ implies $nb^2/D^2 \gg 1$!

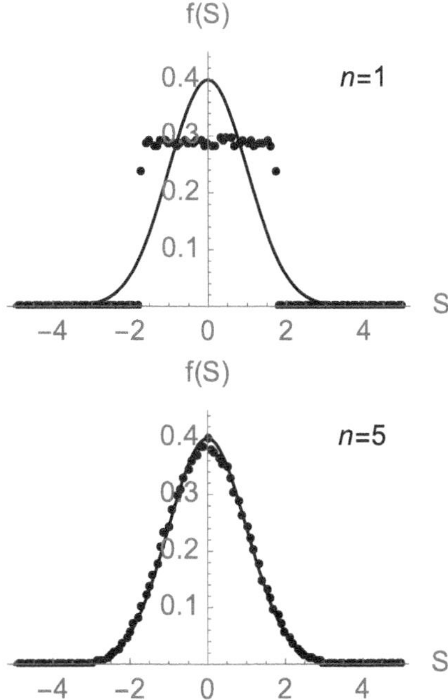

Fig. D.1 Histograms for $n = 1$ and $n = 5$ approximating a standard normal distribution

• *Problem* 2.7 The Zimm plot is depicted in Fig. D.2. The units on the x-axis and on the y-axis are 10^{-14} m^{-2} and 10^{-7} mol/g, respectively. Here the constant c_o is $5 \cdot 10^{13}$ m/kg. The red and the blue lines correspond to the thick lines of the same color in Fig. 2.27. In the present case these lines are quadratic fits through the solid squares. The latter are extrapolations based on the measured data points at constant concentration and constant q, respectively. The corresponding sets of four and five data points are fitted also via quadratic functions, which can then be used to calculate the extrapolated points (solid squares) at zero concentration and zero momentum transfer, respectively. Note that the red and the blue line intersect on the y-axis, as they should, at about 0.68. Hence, we find $\bar{M}_w \approx 1.5 \cdot 10^7$ g/mol. Noda et al. obtain $0.75 \cdot 10^7$ g/mol. However, we should keep in mind that ours is a mere subset of the original data for this sample.

• *Problem* 3.1 (a) We show $\Gamma[x + 1] = x\,\Gamma(x)$ via partial integration applied to $\Gamma[x + 1]$, i.e.

$$\Gamma[x + 1] = \int_0^\infty e^{-t} t^x \, dt \tag{D.19}$$

$$= \Big|_0^\infty (-e^{-t}) t^x - \int_0^\infty (-e^{-t}) t^{x-1} \, dt = x\,\Gamma[x] \; .$$

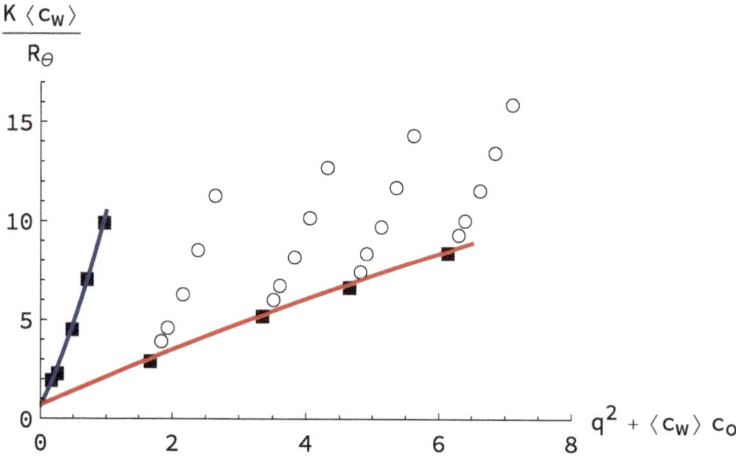

Fig. D.2 Zimm plot of the data in Table 2.6

Since

$$\Gamma[1] = \int_0^\infty e^{-t} dt = 1 \,, \tag{D.20}$$

we can conclude that $N! = \Gamma[N+1]$ is indeed correct.

(b) It is easy to see that $f(t) = \ln t - t/N$, which has a maximum at $t = N$. The expansion yields

$$f(t) = f(N) - \frac{(t-N)^2}{2N^2} + \cdots \,, \tag{D.21}$$

where $f(N) = \ln N - 1$.

(c) Inserting (D.21) into the integral yields

$$N! = \Gamma[N+1] \approx e^{Nf(N)} + \int_{-\infty}^{\infty} \exp\left[-\frac{1}{2}\frac{(t-N)^2}{N}\right]$$

$$= \exp[N \ln N - N] + \sqrt{2\pi N} \,. \tag{D.22}$$

• *Problem* 3.2 (a) Our starting point is the equation of state (3.19), which we expand in small ϕ. The result is

$$\frac{b^3 P}{k_B T} = \frac{\phi}{m} + \frac{1}{2}\left(1 - \frac{\epsilon_o}{k_B T}\right)\phi^2 + \frac{1}{3}\phi^3 + \cdots \,. \tag{D.23}$$

The **Boyle temperature** T_B is the temperature at which the second virial coefficient vanishes. In the present case this means that the term in brackets must be zero, i.e. $T_B = \epsilon_o / k_B$. The critical temperature T_c is given in (3.20). And since (3.24) holds for $m = 1$, we find that it is indeed correct.

(b) First note that instead of $\frac{\partial P}{\partial V}\big|_T = \frac{\partial^2 P}{\partial V^2}\big|_T = 0$ we can use $\frac{\partial P}{\partial \phi}\big|_T = \frac{\partial^2 P}{\partial \phi^2}\big|_T = 0$ (check this!) to compute T_c and ϕ_c. If we apply the latter conditions to P in the expansion (D.23) we find

$$T_c \approx \frac{\epsilon_o}{k_B T} \quad \text{and} \quad \phi_c \approx \frac{1}{\sqrt{m}} \,. \tag{D.24}$$

This works, because ϕ_c tends to zero in the limit of large m. Hence, we do not need the full equation of state.

• *Problem* 3.3 The text should provide you with sufficiently helpful information to do this mainly numerical problem.

• *Problem* 3.4 We assume that the free enthalpy is a function of temperature T, pressure P, and composition, i.e. $G = G(T, P, N_1, \ldots, N_K)$. The quantity K is the number of components, e.g., N_j is the number of molecules of type j in the system. Hence

$$dG = -SdT + VdP + \sum_{i=1}^{K} \mu_i dN_i \,, \tag{D.25}$$

where S is the entropy, V is the volume, and μ_i is the chemical potential per i-molecule. However, we may also use $G = \sum_{i=1}^{N} \mu_i N_i$ to calculate dG:

$$dG = \sum_{i=1}^{K} (\mu_i dN_i + N_i d\mu_i) \,. \tag{D.26}$$

Combination of the two equations and assuming that the temperature is constant yields our starting point—the **Gibbs-Duhem equation** at constant temperature:

$$VdP = \sum_{i=1}^{K} N_i d\mu_i \,. \tag{D.27}$$

Here we deal with a two-component system. Keeping component 2 constant, a pressure increase dP is due to a corresponding increase of the amount of component 1. Integrating the above equation then yields

$$\int_{P}^{P+\Pi_1} VdP' = \int_{0}^{N_1} N_1' d\mu_1(N_1') \tag{D.28}$$

$$= \int_0^{N_1} N_1' \frac{\partial \mu_1(N_1')}{\partial N_1'} dN_1'$$

$$= \int_0^{N_1} N_1' \frac{\partial^2 G}{\partial N_1'^2} dN_1' \, .$$

In a separate short step we rewrite the last integrand:

$$N_1' \frac{\partial^2 G}{\partial N_1'^2} = N_1' \frac{\partial^2 G}{\partial N_1'^2} + \frac{\partial G}{\partial N_1'} - \frac{\partial G}{\partial N_1'} \tag{D.29}$$

$$= -\frac{\partial}{\partial N_1'} \left(-N_1' \frac{\partial G}{\partial N_1'} + G \right) = -\frac{\partial}{\partial N_1'} \frac{\partial G/N_1'}{\partial (1/N_1')} \, .$$

Hence, assuming that the volume does not change in this process, we finally arrive at

$$V \Pi_1 = -\frac{\partial G(N_1)/N_1}{\partial (1/N_1)} + G(N_1 = 0) \, . \tag{D.30}$$

This is almost (3.37). Complete agreement requires two things: First, we identify N_1 with ν_1. Second, here G is the total free enthalpy, but (3.37) contains the mixing free enthalpy ΔG instead. However, inserting $G = G_1 + G_2 + \Delta G$ (note: $G(N_1 = 0) = G_2$) we find that (D.30) indeed agrees with (3.37).

• *Problem* 3.5 (a) We use $c^* \approx (m/b^3)N^{1-3\nu}$ based on (2.124). The molar mass of the monomer is about $m \approx 118$ g/mol. Here we simply approximate b^3 by the monomer volume (which of course is rough). The latter follows via the bulk density of the polymer of about 910 kg/m^3, i.e. $b^3 \approx 130$ cm^3 per mol. Using our results from problem 2 we also have $N \approx 590$ (α-104), 1600 (α-12), and 4300 (α-103). For $\nu = 1/2$ this yields $c^*_{\alpha-104} \approx 3.7 \cdot 10^{-2}$ g/cm^3, $c^*_{\alpha-12} \approx 2.3 \cdot 10^{-2}$ g/cm^3, and $c^*_{\alpha-103} \approx 1.4 \cdot 10^{-2}$ g/cm^3; for $\nu = 3/5$ we find $c^*_{\alpha-104} \approx 0.55 \cdot 10^{-2}$ g/cm^3, $c^*_{\alpha-12} \approx 0.25 \cdot 10^{-2}$ g/cm^3, and $c^*_{\alpha-103} \approx 0.11 \cdot 10^{-2}$ g/cm^3 instead.

(b) Using $Z(\rho/\rho^*) \sim (\rho/\rho^*)^x$ in (3.64) yields

$$\Pi \sim \rho^{x+1} N^{(3\nu-1)x-1} \, , \tag{D.31}$$

and therefore $(3\nu - 1)x - 1 = 0$ if we want to satisfy the condition regarding N. Hence $x = 2$ if $\nu = 1/2$ and $x = 1.25$ if $\nu = 3/5$.

(c) The plot is shown in Fig. D.3. Note that we can combine (3.63) with (3.64) to describe the data in the entire concentration range (dashed line).

Remark 1: You should try and use $x = 2$, the proper value if the experiment is conducted under θ-conditions. Since it is not, $x = 2$ does not work here.

Remark 2: But what is the relation of our scaling theory to the previously calculated mean field osmotic pressure (3.38) obtained via a lattice model? The lattice model

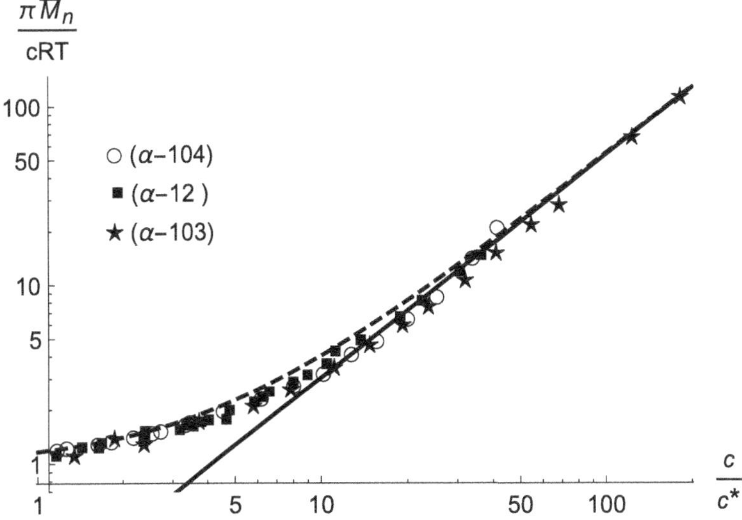

Fig. D.3 Scaling plot of the indicated three data sets from Fig. 1.6. The solid line is $A(c/c^*)^{1.25}$, where A is an adjustable constant. The dashed line is the function $1 + A(c/c^*)^{1.25}$

also yields an N-independent osmotic pressure if we look past van't Hoff's law. But the functional form is not right—except in one special case. This is if we define, as before, θ-conditions via a vanishing second virial coefficient. This means that the leading contribution to the osmotic pressure becomes proportional to ϕ^3 (or c^3) (cf. (3.39)). θ-conditions also mean that the chains are ideal, i.e $\nu = 1/2$ and thus $x = 2$. In this case we find that according to (3.64) the osmotic pressure obtained via scaling theory also follows a c^3-behavior.

Remark 3: We have approximated the quantity b^3 by the monomer volume. Had we chosen to use a better approximation, what would this at have done to Fig. D.3? It would have shifted the $c/c*$-axis.

- *Problem* 4.1 (a)

$$\frac{U}{N} = \frac{1}{N}\sum_{i<j}^{N} u_{ij} \tag{D.32}$$

$$\approx \frac{4\pi}{2}\rho \int_r^\infty dr' r'^2 u(r') = \frac{8\pi}{9}\epsilon\sigma^6\rho^2(\sigma^6\rho^2 - 3)\,.$$

Here we use that there are $N(N-1)/2 \approx N^2/2$ pair interactions between the particles. In addition, $\rho = N/V$ and $u(r)$ is the Lennard-Jones potential.

(b)

$$\rho_o = \sqrt{\frac{3}{2}} \frac{1}{\sigma^3} \, . \tag{D.33}$$

(c) K follows via $K = -V \partial P / \partial V = V \partial^2 U / \partial V^2$ or

$$K = \rho \frac{\partial}{\partial \rho} \rho^2 \frac{\partial (U/N)}{\partial \rho} \, . \tag{D.34}$$

Inserting U/N from (D.32) and replacing ρ by its equilibrium value in (D.33) yields

$$K = \frac{8\sqrt{6}\pi\epsilon}{\sigma} \approx 62 \frac{\epsilon}{\sigma^3} \, . \tag{D.35}$$

(d) You should obtain $K \approx 2.3$ GPa, which is quite close to the values in the reference. This is not too surprising, since elastomers behave more or less like liquids when compressed and the Lennard-Jones potential models liquids quite well.

• *Problem* 4.2 (a) We apply formula (4.20), i.e. $\sigma_{zz} = \frac{N}{V} k_B T \left(\lambda - \frac{1}{\lambda^2} \right)$, where N/V is the segment number density. Here we use the linear fit through the data for $\lambda = 1.383$. Its slope, $\Delta\sigma_{zz}/\Delta T$, is approximately 400 N m^{-2} K^{-1}. The slope of the bottom line obtained with $\lambda = 1.043$ is -100 N m^{-2}. We assume that this slope is due mainly to thermal expansion and, correcting for thermal expansion, we conclude that the slope due to entropy elasticity for $\lambda = 1.383$ is $\Delta\sigma_{zz}/\Delta T \approx 500$ N m^{-2} K^{-1}. Note that we use the data obtained for $\lambda = 1.383$ rather than for instance the data for $\lambda = 1.961$. This is because the validity of the formula for σ_{zz} degrades quickly when λ is large—but clearly, here we are not on solid ground. That said, N/V is straightforwardly obtained, i.e. $N/V \approx 4.2 \cdot 10^{25}$ m^{-3}. A monomer of natural rubber, containing $5\times$C and $8\times$H, is depicted in Fig. 1.2 (cis-1,4-) and hence its molar mass is 68 g/mol. This means that one cubic metre of NR contains about $8.1 \cdot 10^{27}$ monomers. Dividing this number by N/V we obtain an average segment length of 190 monomers.

(b) Addition of 1.5 % (by weight) of dicumyl peroxide means about $3 \cdot 10^{25}$ molecules per cubic metre. If we divide the above number for the monomers per cubic metre by this number, we obtain 260 monomers per dicumyl peroxide molecule. This, however, is not the segment length. Note that a cross-link is the junction of four segments. This means that we have two segments per cross-link (think about this!) and therefore the segment length according to this estimate is only 130 monomers. Considering the very rough nature of these estimates, the overall comparison is not too bad (see also the remark at the end of Problem 5.1!). However, bear in mind that our first estimate also includes the physical cross-links, whereas the second does not. On the other hand, the second estimate is based on the assumption that each dicumyl peroxide molecule produces a cross-link—which is not really true.

• *Problem* 4.3 (a) In the case of steady shear (4.76) becomes

$$\sigma(t) = \cdot\gamma \int_{-\infty}^{t} G(t-t')dt' = \cdot\gamma \int_{0}^{\infty} G(s)ds . \tag{D.36}$$

Here we have used the substitution $s = t - t'$. Comparing this to (4.49) yields (4.218).

(b) In the context of (4.135) and (4.136) we had concluded that every term in the summations corresponds to a Maxwell model, which in turn implies that the shear relaxation modulus of the Maxwell model is $G_M(t) = G_M(0)\exp[-t/\tau_M]$. Inserting $G_M(t)$ for $G(t)$ in (4.218) yields again (A.6).

• *Problem* 4.4 (a) Insertion of $G(t) \sim A\exp[-t/\tau]$ into (4.218) yields $\eta \sim A\tau$. This justifies (4.220), since $G(\tau) \sim A$.

(b) Following the motivation used to justify (4.131) we write

$$G(\tau) \sim k_B T \rho . \tag{D.37}$$

With $\rho = \phi/(Nb^3)$, where ϕ is the volume fraction polymer and $c = m_b\phi/b^3$, where m_b is the mass of a monomer (or alternatively one Kuhn segment), we find

$$\eta_i \sim \frac{k_B T}{\eta_s m_b N}\tau . \tag{D.38}$$

(c) We use (4.126), a result of the Rouse model, to eliminate D_{cm}. In addition, $R \approx bN^\nu$. Hence,

$$\tau_R \sim \frac{\eta_s b^3}{k_B T}N^{2\nu+1} . \tag{D.39}$$

Note that $\zeta \sim \eta_s b$ (Stoke's friction).

(d) Combination of (D.38) with (D.39), using $\tau \sim \tau_R$, yields

$$\eta_i \sim \frac{b^3}{m_b}N^{2\nu} \quad \text{(Rouse)} . \tag{D.40}$$

(e) Instead of the Rouse result for D_{cm}, i.e. (4.126), we now use the Zimm result, i.e. $D_{cm} \propto N^{-\nu}$ (cf. (4.165)). Hence

$$\eta_i \sim \frac{b^3}{m_b}N^{3\nu-1} \quad \text{(Zimm)} . \tag{D.41}$$

Remark: The relation between viscosity and molecular weight is a topic whose research predates the models of Rouse and Zimm. For the interested reader I recommend [2, 3] by Fox and Flory. The empirical equation

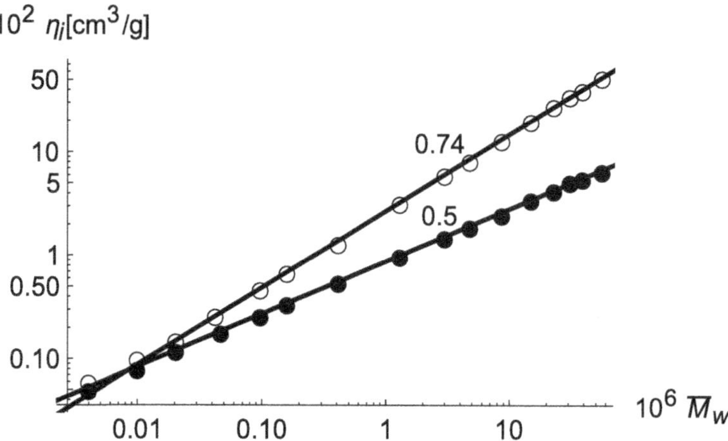

Fig. D.4 Intrinsic viscosities vs. weight-average molar mass for polystyrene in benzene (open circles) and cyclohexane (solid circles). The data are from Table I in [1]. The numbers are the values obtained for the parameter a, when the data are fitted with the Mark-Houwink equation

$$\eta_i = K M^a \, , \tag{D.42}$$

where K and a are parameters (tabulated a-values based on fitted data range from 0.5 (θ-conditions) to about 0.8) and M is the polymer molecular weight, is known as **Mark-Houwink equation**.

(f) The plot is depicted in Fig. D.4. Note that $a = 3\nu - 1 = 0.5$ means $\nu = 1/2$ (θ-conditions) in (D.41). The exponent $3\nu - 1$ evaluated in the case of a good solvent, using the Flory value 3/5 for ν, yields $a = 0.8$. This value is not really far from 0.74. However, if we use the Rouse prediction instead we obtain $a = 1$ and 1.2, which obviously is not supported by the data.

• *Problem* 4.5 (a) Note that σ_M satisfies (A.2) with $G = G_2$ and $\gamma_M = \gamma$. Inserting $\sigma_M = \sigma - \sigma_1 = \sigma - G_1\gamma$ into this equation yields (A.7).

(b) This should be straightforward to calculate.

• *Problem* 5.1 (a) In (4.20) we start with $\sigma_{zz} = \frac{1}{V} \frac{\partial(-T \Delta S^{el}_{stretch})}{\partial \lambda}$, which is a bit hasty. It is better to start from $f_z = \frac{\partial(-T \Delta S^{el}_{stretch})}{\partial L_z}$, where f_z is the force exerted by the apparatus stretching the sample. The true stress σ^t_{zz} is given by

$$\sigma^t_{zz} = \frac{f_z}{L_x L_y} = \frac{1}{L_x L_y} \frac{\partial(-T \Delta S^{el}_{stretch})}{\partial L_z} \, . \tag{D.43}$$

Here L_x, L_y, and L_z are the instantaneous dimensions of the sample, whose original dimensions are $L_{x,o}$, $L_{y,o}$, and $L_{z,o}$. Volume conservation means $L_x = L_{x,o}/\sqrt{\lambda}$, $L_y = L_{y,o}/\sqrt{\lambda}$, and $L_z = L_{z,o}\lambda$, i.e. $V = L_x L_y L_z = L_{x,o}L_{y,o}L_{z,o}$. Inserting this into the above equation yields

$$\sigma_{zz}^t = \frac{\lambda}{V} \frac{\partial(-T\Delta S_{stretch}^{el})}{\partial\lambda} .$$

(D.44)

Since for uniaxial deformation at constant volume $\sigma_{zz}^t = \lambda\sigma_{zz}^e$, where σ_{zz}^e is the engineering stress, we conclude that σ_{zz} in (4.20) is the engineering stress.

(b) The **elastic stress tensor**, introduced in (4.10), is computed via the instantaneous local strain tensor components, i.e. it is the true stress. However, our comparison of (4.19), based on the theory of elasticity, with (4.21), based on chain entropy, is in the limit of small strain ϵ (note: $\lambda = 1 + \epsilon$). Since the expansions of $\sigma_{zz}^t = \sigma_o(\lambda^2 - 1/\lambda) = \sigma_o(3\epsilon + \epsilon^3 + \cdots)$ and $\sigma_{zz}^e = \sigma_o(\lambda - 1/\lambda^2) = \sigma_o(3\epsilon - 3\epsilon^2 + \cdots)$, where $\sigma_o = (N/V)k_BT$, agree to first order in ϵ, there is no difference between σ_{zz}^t and σ_{zz}^e with respect to the comparison with the continuum elasticity result.

Remark: In Problem 4.2 we assume that the experimental stress is the engineering stress. However, using the true stress instead improves the comparison between the segment lengths in parts (a) and (b).

• *Problem* 5.2 (a) Use spherical coordinates and carry out the angular integration analogous to the one in (2.99). The subsequent radial integration yields

$$f(q)/Q = \frac{3}{(qR)^3}(\sin[qR] - qR\cos[qR]) .$$

(D.45)

$f^2(q)$ is the solid red line in Fig. 5.40, where Q is adjusted to match the experimental intensity. Note that in A. Bouty et al. Q is our q and their form factor is $f^2(q)$ instead of $f(q)$.

(b) Obviously, the oscillations in $f^2(q)$ are not observed experimentally. The main reason is the polydispersity of the particle population. Here we mimic this polydispersity using a normalized mock particle distribution described by $g(R)$ depicted in Fig. D.5. Its specific form, i.e. $R^6 \exp[-(R/R_o)^2]$, is motivated by mathematical convenience (in the above paper the authors use a log-normal distribution instead). Note that the average particle size $\bar{R}$, computed via $g(R)$, is $16\,R_o/(5\sqrt{\pi}) \approx 1.8\,R_o$.

Fig. D.5 Mock particle size distribution function $g(R)$

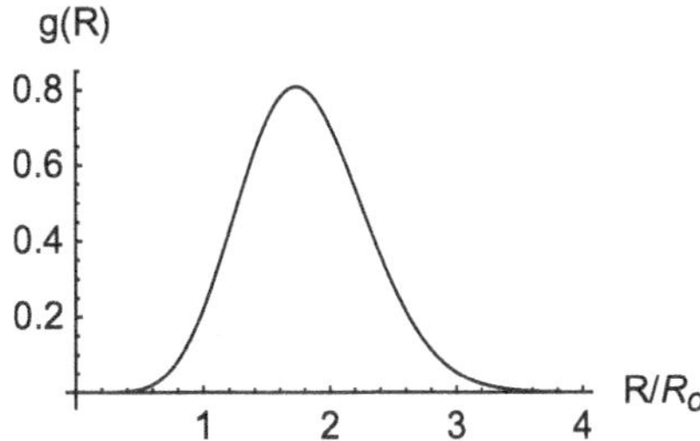

The integration, i.e. $\bar{f}^2(q) = \int_0^\infty f^2(q)g(R)dR$, yields

$$\bar{f}^2(q)/Q^2 = \frac{6}{5x^4} + \frac{12}{5x^6} - \left(\frac{12}{5x^2} + \frac{18}{5x^4} + \frac{12}{5x^6}\right)e^{-x^2}, \qquad (D.46)$$

where $x = qR_o$. We can see that $\bar{f}^2(q)/Q^2 \propto x^{-4}$ in the limit of large x. If x is small, the leading behavior is $\bar{f}^2(q)/Q^2 \approx 1 - 7x^2/10$. It is this $\bar{f}^2(q)$, again with Q adjusted to match the experimental intensity, which is shown as a solid black line in Fig. 5.40. Note that we use $R_o = 44$ Å to plot the experimental data. This is because in the paper the authors state that the average primary particle size is roughly 80 Å.

Remark: For particles possessing a fractal surface structure, characterised by a surface fractal exponent d_s, one can show that Porod's law is replaced by q^{-6+d_s} [4].

(c) It is straightforward to check the power-laws included in Fig. 5.40. However, it is worth mentioning their physical significance. The power-law with the apparent exponent $d_f \approx 2.3$ is attributed to the fractal nature of small aggregates formed by primary particles. Since the aggregates are larger than the primary particles, they contribute to the scattering intensity when $q > R_o^{-1}$. The subsequent exponent, $d_f \approx 1.5$, is attributed to the formation of a likewise fractal filler structure, perhaps sections of a beginning network, on a still larger scale. However, the respective q-intervals in which these power-laws are observed are not wide, perhaps one order of magnitude in the case of $d_f \approx 2.3$ and maybe two orders of magnitude in the case of $d_f \approx 1.5$. This means that one must be careful with any interpretation of the apparent structures.

References

1. T.G. Fox, P.J. Flory, Viscosity-molecular weight and viscosity-temperature relationships for polystyrene and polyisobutylene. J. Am. Chem. Soc. **70**, 2384 (1948)
2. T.G. Fox, P.J. Flory, Second-order transition temperatures and related properties of polystyrene. I. Influence of molecular weight. J. Appl. Phys. **21**, 581 (1950)
3. Y. Einaga, Y. Miyaki, H. Fujita, Intrinsic viscosity of polystyrene. J. Polym. Sci., Polym. Phys. **17**, 2103 (1979)
4. H.D. Bale, P. Schmidt, Small-angle X-ray-scattering investigation of submicroscopic porosity with fractal properties. Phys. Rev. Lett. **53**, 596 (1984)

Index

R. Hentschke, *A Concise Introduction to Polymer Physics,*, Undergraduate Lecture Notes
in Physics, https://doi.org/10.1007/978-3-031-87324-9